TIME:
A PHILOSOPHICAL
TREATMENT

TIME

A PHILOSOPHICAL TREATMENT

KEITH SEDDON

CROOM HELM
London • New York • Sydney

Croom Helm Ltd, Provident House, Burrell Row,
Beckenham, Kent BR3 1AT
Croom Helm Australia, 44–50 Waterloo Road,
North Ryde, 2113, New South Wales

British Library Cataloguing in Publication Data

Seddon, Keith
Time: a philosophical treatment
1. Time
I. Title
115 QB209

ISBN 0–7099–5424–7

Published in the USA by
Croom Helm
in association with Methuen, Inc.
29 West 35th Street
New York, NY 10001

Library of Congress Cataloging-in-Publication Data

Seddon, Keith, 1956-
Time: a philosophical treatment.

Bibliography: p.
Includes index.
1. Time. I. Title.
BD638.S43 1987 115 87-3607
ISBN 0-7099-5424-7

Printed and bound in Great Britain by Mackays of Chatham Ltd, Kent

Contents

Part One

1
Introductory — The Dynamic View of Time

The strange thing about time is that the way people ordinarily think of it is completely wrong — so I intend to show. If we pretend to have access to the secret thoughts of a person of common sense who is thinking about where they have come to in life and where they hope to go we will see the way in which time is mistakenly conceived. This imaginary person is like each and every one of us. Until we do philosophy and start examining the way we think, we all think along very similar lines. All that I have done, my successes and failures, lie in the past, thinks this person. That my failures are receding into the past is not such a bad thing, but how sad it is that the successes, the enjoyable times, must go as well. But not to worry, for there is the future, with its fresh adventures, advancing towards me. All the things that I shall do lie here. Is it not a consoling thought that successes yet to be mine are even now drawing nearer and nearer, until that magic moment comes when they will exist in the present, and I will enjoy what all men take delight in when what I have wanted and planned for will be with me?

We speak this way about our lives and our endeavours all the time. The philosophical interest in this comes when we wonder whether our talk says anything metaphysically correct about time and events. For what we seem to be saying is that time flows, or moves, such that events are constantly changing their position in relation to the present moment: or else it is we who are steadily advancing into the future, experiencing the events in our lives which lie there as we go. We speak of events as though they are 'dynamic' in the one sense as moving through time, either towards us from the future or away from us into the past, and in the second sense as changing their temporal positions. Both senses are appealed to when it is said that future events become less and less future until they become present, whereafter they pass into the past, and thence recede further and further into the past. Along with the events that change their position relative to the present come the dates at which they occur: 1 January 2001 for instance is constantly drawing closer to us. Soon enough that date will be present. Then it will be a date in recent history; it will move further and further away from the present, deeper and deeper into the past. Eventually it will be a date in ancient history, and our achievements in this present era will have passed away into the dark recesses of the past.

Physical objects are subject to a similar movement through time. Of objects that are no longer with us (such as the original St Paul's Cathedral which was destroyed in the Great Fire of London) it seems natural to say that they have passed into the past, and are receding forever from us along with all past dates and events.

George N. Schlesinger claims that it is a fact of human experience that events flow towards us from the future, are experienced in the present, whereupon they flow on again into the past.[1] Those who adopt this view would point to any number of ordinary-language expressions which reflect our apparent belief that time flows. People say that 'time flies', that 'time rushes by', that the crisis of an illness 'is approaching', and 'the world has passed by' someone; we talk of 'the river of time', of 'advancing through time'; 'it's all water under the bridge now'; 'time has slipped away'; 'tomorrow is still to come'; we 'while away' an hour; 'time creeps by'; 'time marches on'; 'we've lost time'; 'the clock has lost time'; 'I think I can find enough time'; 'where has the time gone?'; time is wasted and spent; sometimes there is 'no more time' (it has all been used up); 'I have plenty of time on my hands'.

In ordinary language, time is discussed in terms of metaphors, many of which bring with them the image of time flowing like a liquid, of time being a sort of stuff that can be stopped up, or spent, or used carelessly. 'Tomorrow is coming, moves in, moves on, is gone, joins yesterday. It will never come by this way again. Time does not stand still. Nor does tomorrow come in, move out, and then rest. It keeps on going and every day it's further away.'[2] The adherent of the dynamic view of time takes these expressions not really as metaphors at all. They express for him the honest truth about reality: time flows.

The truth is that we think of time on the model of a flowing river (or perhaps a moving conveyor belt, or a speeding locomotive, or some other image of movement). We are like passengers in a boat drifting down the river. The scenes that pass us are the events in our lives. Behind us, receding upstream, are all our past experiences, now for ever beyond our reach. We cannot even see them any more; they exist for us only in memory. And downstream, ahead of us, lies the future, steadily getting closer and closer. Or if we prefer, we can dispense with the boat and see ourselves standing on a bridge or sitting on a bank beside the river. The twigs and leaves and boats that float past us represent the events that we experience. Upstream is our future, flowing towards us. Downstream is our past.[3]

Someone, like Schlesinger, who takes this river-of-time model seriously and believes that events really do move through time (in either sense that they pass us, or we pass them), I shall refer to as holding the transient view of time. Time, on this view, is fully dynamic in the two senses already mentioned. Firstly, temporal motion is an objective fact about reality, and secondly, events really do change with respect to being past, present or future (thus, an event now present was once future, and will later be past). It is possible to deny that events are dynamic in the first sense, that is, events do not really move, but still hold that events are dynamic in the second sense, that is, they do change with respect to being past, present and future. Someone who thinks this I will refer to as holding the tensed view of time (it will be clear later why 'tensed view' is a suitable expression). Thus, someone may believe that events really do change with respect to their being past, present or future, yet not cash this change in terms of events moving through time from the future, to the present, and on into the past. Someone who adheres to what I am calling the tensed view of time would find the river-of-time model suspect, and nothing other than a convenient but metaphysically misleading image. This being the case, it does not follow that events do not change in respect to being past, present and future. Thus, a transient theorist would maintain and a tensed theorist deny that our talk about the flow of time and the movement of events is a reliable guide to the claim that events move through time. Both theorists take such talk to be a reliable guide to the claim that events change with respect to being past, present and future. One could deny that events are transient, in the sense that they really move through time, but not deny that events change with respect to being past, present and future. One could not deny that events change with respect to being past, present and future without also denying that events really move through time. If an event is to move, it must start off being future so that it can move to a temporal position where it is less future, or it must start off being present so that it can move into the past, or it must start off past so that it can move further into the past. If an event cannot change with respect to its being past, present or future, it cannot move in the way the transient theorist holds that all events move.

In Part I of this discussion, I will pause to look more closely at the ways in which our ordinary-language expressions confuse us about the nature of time; I then intend to show that the notion of temporal movement as incorporated into the transient view is incoherent along with a conception of 'the NOW' which Schlesinger employs to defend the transient view. I will then show why I think the tensed view of time is mistaken, arguing that events do not really change with respect to being past, present and future — that events logically resemble objects in space which are not intrinsically 'here' and 'there' but simply related spatially to each other; similarly, events are related temporally to each other, any event being earlier than some other events, and later than some other events, but no event is really past or present or future. I will show what is required for a tensed statement to be true ('E is future' is a tensed statement — these terms will be explained in the proper place), and I will show why it is that no tensed statement can be translated by a tenseless statement (a statement which says how events are related temporally to each other making no reference to the present). I will also make mention of McTaggart's remarks about time and change, disputing with him his claim that change is impossible if time is not tensed (that is, if events do not change with respect to being past, present and future).

Notes

1. 'How Time Flies', which is essentially the same material as Chapter 4 of his book *Metaphysics*. (Full details to references are given in the Bibliography.)

2. O. K. Bouwsma, 'The Mystery of Time (or, The Man Who Did Not Know What Time Is)'.

3. See Donald C. Williams, 'The Myth of Passage' pp. 103 ff. for more on temporal imagery.

2
The Static View of Time

If we wish to deny that time is dynamic and reject the transient and the tensed views of time, we can do this by maintaining the static view of time. On this view, events are ordered by the relation 'earlier than' (or its logical opposite 'later than'); and that events are so ordered is not cashed in terms of the sequence in which events cease being future and become present, and is not cashed in terms of events moving in time. The static view of time holds that there is no moving present, and there is no flow of time. Such notions are simply mistaken. The river-of-time image, although seemingly indispensable when we need to think about ourselves, our lives, our plans for the future and our recollections of what has been, is a fraud, having nothing useful to instruct us about the true nature of time and events. Why we have this image and how it misleads our thought about time will be discussed in the next section.

Events are not intrinsically past, present and future, and they do not change in respect of being past, present and future, despite the fact that we speak of them as if they do. Time is just a matter of relation between events. If event E_1 occurs earlier than event E_2 we express all there is to say about the temporal state of affairs concerning E_1 and E_2 by stating that E_1 is earlier than E_2. This fact does not consist in any further facts, such as E_1 being future while E_2 is even more future, or E_1 being past while E_2 is future, or E_2 being past and E_1 being even more past; neither does it consist in the fact that E_1 attains presentness before E_2 does. The static view of time denies that there is a present at all, in which case events do not become present. Even though we experience events 'in the present' and talk of experiencing events 'in the present' the events we refer to when saying these things do not have something which all other events, past and future, lack. Present events have not gained something which future events have yet to acquire, and past events have not lost something which they once had. This view is appropriately called the 'static' view, because it denies that time is dynamic in the two senses already mentioned; that is, it denies that events move from future to past or that we move towards the future, and it denies that events change with respect to being past, present and future.

My aim throughout this discussion will be to object to thinking about time dynamically, and to show that the static view can withstand the objections from the transient and tensed theorists, and constitutes an adequate theory of time.

It ought to be noted before moving on that the temporal relation 'earlier than' is transitive and asymmetric. By transitive we mean that if E_1 is earlier than E_2 and E_2 is earlier than E_3, E_1 is earlier than E_3. In general terms we can say that if one particular bears the relation concerned to another particular, and that particular bears the relation to a third particular, then the first particular bears that relation to the third particular. If we experience E_1 before E_2, and E_2 before E_3, there is nothing further to experience or find out in order to claim correctly that E_1 is earlier than E_3. Another example of a transitive relation would be 'heavier than'. The relation 'owes money to' is not transitive. And by 'asymmetric' we mean that if E_1 is earlier than E_2, then E_2 logically cannot be earlier than E_1. This is one of those relations, which in general terms, is such that if one particular bears it towards another, that other particular cannot bear it to the first. 'Taller than' is similarly asymmetric. This understanding of 'earlier than' would be disputed by people who think that time is cyclic, that the whole history of the universe, having happened, starts again at the beginning and happens again, and so on. I shall not have occasion to address this strange idea.

3

Ordinary Language and the Nature of Time

J. J. C. Smart remarks that 'certainly we *feel* that time flows', but this feeling he believes 'arises out of metaphysical confusion',[1] which is what I believe and what I hope to elucidate in the course of this discussion. We feel that time flows because from the beginning when we were small children our acquaintance with time, with things happening, having happened and about to happen was mediated by metaphors of movement and flowing (some of which were noted in Chapter 1). I remember it very well. Events yet to happen were said to be approaching me: the school holidays were approaching, at another time the new school term was approaching; the time for the bandages to be removed from my injured thumb was approaching. I remember being asked at the age of five whether I was looking forward to starting school. I knew a little bit of what went on in school, and when I was asked that question I found myself picturing an image of a classroom with myself in the room — here was a picture of the future, fast approaching; it was like seeing a scene further down a road which when I arrived at that spot I would be involved in. Everywhere there are diaries, calendars and wall-charts, representing time as a ribbon along which we travel. Whenever anyone asks 'What are you doing next week?' I see in my mind's eye that temporal ribbon, neatly divided off into separate days, by means of which I can recall my plans. Part of the answer as to why we feel time flows obviously lies in the fact that to talk about time at all involves talking with spatial metaphors and movement metaphors. The language we have all grown up with dictates a pattern of thought. Even though I now feel convinced that time does not flow, if I think about the past or the future I can do this only by thinking in terms of the river-of-time image; without the image I could not think about time at all. Since I cannot dispense with the image, all I can do is remind myself that as far as the metaphysical truth about time is concerned, the image is false.

There are three points I want to mention which contribute to an understanding of why we have the sort of temporal language which we do. These are tentative suggestions, and do not say the last word on this difficult question.

Human experience is comprised of a ceaselessly changing panorama of events, what Broad calls 'that series of successive experiences which constitutes one's mental history from the cradle to the grave'.[2] We are experiencing different things all the time. Even for the man locked away in solitary confinement, his experience will be that of one thought, one image, after another, of different bodily sensations, of successive distant noises. Seeing time in terms of a flowing river seems to be an attempt to explain why it is that our experience is indeed a ceaselessly changing panorama. Here we sit beside the river of time, and the events we experience are brought to us upon the never-ending current. A fixed pattern of events floating upon the river produces for us our ever-changing panorama as the events drift by.

People can remember what has happened, and they can anticipate what is yet to come. I suspect there is a tendency to model remembering and anticipating on ordinary perception. For many people (certainly for me), recalling a past experience very frequently involves having a visual image, and to this extent resembles ordinary seeing. The mistake that can be made is to think of what is remembered or what is anticipated as having a real existence somewhere or other, just as an object perceived is regarded as having a real existence in a perfectly straight forward sense. The difficulty is to understand how a past or future event can have an existence which is different from the existence of an event which is perceived in the present. Thinking in terms of the river of time offers a solution. Present, past, and future events all have essentially the same sort of existence; it's just that those events which are now flowing past us, which we call present, can be directly perceived, whereas past events, still existing in their own right, can only be remembered, and future events can only be anticipated. (We should note that the relation 'downstream of', like 'earlier than', is transitive and asymmetric, making the river image all the easier to assume.) Clearly there is not much of a philosophical theory here. These ideas about the river of time seem circular and arbitrary. But in so far as the common man thinks about time at all, this, I feel, is the way in which it is done.

Thirdly, and lastly, given that people remember what has happened to them, our experience is that we are forever accumulating a greater and greater stock of memories. We are being filled up by our experiences. People talk of gaining experiences as much as simply having them. That this is how we find our lives makes it easy to 'hypostatise' time (as Smart has put it[3]); our memories are like vessels into which our ever-changing experiences are poured. It is true that we accumulate, via the faculty of memory, more and more experiences, and the only way we have of representing accumulation to ourselves is to think in terms of vessels being filled.[4]

Schlesinger is very impressed by the fact that 'human beings in widely different cultural settings and in all periods of history have regarded it as one of the most central features of existence that time moves, so that events are carried from the future towards us and then recede further and further into the past'.[5] He does not say, but I take it that he knows this to be the case by simply noting the way in which people have talked about time. They talked in terms of models such as the river of time, revealing a belief that time moves. This can be conceded. But the observation that people talk as if time flows is no guide to the philosophical truth of the matter. With respect to a deity or deities, we can say again that 'human beings in widely different cultural settings and in all periods of history have regarded it as one of the most central features of existence' that there exists a deity or deities. This claim is probably true, but still leaves undecided the philosophical question as to whether a deity or deities exist. My belief is that Schlesinger sees a wrong importance in the fact that people talk about time as if it moves. For him it indicates that time moves, for me it indicates that Schlesinger, and others, have been too easily taken in by the surface appearance of our language. It is not difficult to see how our temporal language leads us to have mistaken ideas about the nature of time.

In our language we can find groups of sentences which have the same 'surface grammar', the same 'form of expression', but which have a different 'depth grammar'. One of Wittgenstein's main contentions was that philosophical confusion arises in many areas because people have failed to notice this fact of language, and have been misled by similarities of surface grammar; a form of expression misleads because we assimilate it with another expression which has the same surface grammar. Henry Le Roy Finch sums this up neatly with several apt references to Wittgenstein in his paragraph:

> On its surface, grammar is full of similes which create false appearances (PI 112)[6] and pictures which 'force themselves on us' (PI 140, 397) and 'hold us captive' (PI 115). It fascinates us with misleading analogies ([*Blue Book* page] 49) and tempts us to misunderstanding and invent myths (PI 109, 345). We try to follow up the analogies which it suggests and we find that they conflict with each other and we get entangled in our own rules (PI 125). Grammar is a snare and a delusion.[7]

Time is particularly prone to this difficulty. Sometimes we speak of the past as if it is a place. The instructions 'Don't live in London' and 'Don't live in the past' have the same surface grammar and give the impression that the past is a location in the way that London is. This is because 'live in' has a different depth grammar in either statement. In the first, 'live in' refers to someone's physical location, but in the second statement it refers to the having of a certain outlook or attitude to life. Similarly, the statements 'I'm putting this book in the bookcase' and 'I'm putting that bad experience behind me' make it look as though putting a bad experience behind one is essentially the same sort of action as putting a book in a bookcase. We know that this is not the case, since the past is not really a place where past experiences, good or bad, have their locations. But the damage is already done. We talk in these terms, and the image of the past as a location has been conjured. Other expressions we use (and there are scores of them) create the false picture of time as a thing which moves. Just as we say 'The procession is approaching' we say 'The exams are approaching.' The surface grammar of these two statements is the same, but the depth grammar is different because 'approach' has got more than one meaning. With regard to any set of statements which share the same surface grammar, we can determine whether they have also the same depth grammar by seeing whether the assumptions which can be properly held and the questions which are appropriate to ask carry over from any particular sample in the set to the other statements. When we look at the statement 'Don't live in London' we can see that it is right to assume that London is a real place, and one can appropriately ask how one may get to

London. But when we look at the statement 'Don't live in the past' we can see that 'the past' is not meant to refer to a real location, and it would be absurd to ask how one might get there.

Wittgenstein addresses himself to the topic of time in section I.56 of 'The Brown Book',[8] emphasising how the grammatical form of statements dealing with non-temporal subject matters can confuse us. He looks at the question 'where does the present go when it becomes past, and where is the past?' He is very impatient with this question, saying that 'we should wave it away as non-sense'. The question can be asked because we can ask analogous questions, with the same surface grammar, in other contexts. Wittgenstein mentions just one, that of logs floating down a river. 'In such a case we can say the logs which *have passed* us are all down towards the left and the logs which *will pass* us are up towards the right.' We adopt this simile into our talk about time and events, and thence are derived the expressions some of which we noted in Chapter 1. So just as a log passes by, we speak of present events passing by. In which case, since the question 'Where has the log gone?' is coherent and has a clear answer which is in principle verifiable (for instance, 'Downstream to the left') we are led to expect the question 'Where has that past event gone?' to have an analogous answer, and the invention of 'the river of time' is an attempt to construct a scenario for such an answer. And the answer we get is 'The past event is receding into the past.' The fact that there is an answer at all makes it look as if the original question was asking something sensible. The mistake is to see the answer as expressing a philosophical theory about time, namely that there is a realm of past events where present events go to when they stop being present (and similarly that there is a realm of future events where events that are not yet present are awaiting their moment of glory). When we see that, and when we see that 'flow' has not just one use in all contexts, but several different uses in different contexts, we see that there never was a problem about the flow of time. (This treatment expands somewhat on Wittgenstein's remarks. But I have not said anything with which he would disagree.)

Finally, on this topic of our ordinary talk about time, it is important to note that the word 'time' in English is a noun, and there is a natural tendency to see nouns as the names of objectively existing entities. It is too easy to assimilate the question 'What is the nature of time?' with questions like 'What is the nature of water?' In seeing time on the river model, people are trying to make time into a something-or-other which has a set of qualities, one of those being that it flows or moves. 'We are up against one of the great sources of philosophical bewilderment,' warns Wittgenstein. 'A substantive makes us look for a thing that corresponds to it.'[9] But time is not a thing like this in any sense at all. The expression 'time' is used in many different ways in many different contexts. There is no one 'thing' that is being denoted when the term is used. Even though one may say 'I have plenty of time on my hands', it is not the case that time can stick to hands in the way jam can. Expressions like these make time look like a commodity, when it is not.

Notes

1. 'Time and Becoming'.
2. 'Ostensible Temporality', p. 138.
3. 'The River of Time', p. 216.
4. cf. Smart, 'Time and Becoming', p. 12.
5. *Metaphysics*, p. 100.
6. i.e., *Philosophical Investigations*, paragraph 112.
7. *Wittgenstein — The Later Philosophy*, p. 161.
8. *The Blue and Brown Books*, pp. 107–9.
9. *The Blue and Brown Books*, p. 1.

4
Temporal Movement Denied

Having suggested in the previous chapter that our tendency to think that time flows or events move arises from ignoring the fact that similar surface grammar does not always indicate similarity of depth grammar, I want in this chapter to show that the idea of temporal movement is conceptually incoherent.

When this is done, the dynamic view of time will have suffered injury to at least this extent: if events in time really do change with respect to being past, present and future, that change is not accompanied by, nor mediated by in any sense, any genuine movement on the part of events. Some readers may feel that the notion that time flows has absolutely no plausibility and wonder why it needs to be treated at all. I sympathise with G. E. Moore's attitude to 'absurd' philosophical theories where he says:

> I confess I feel that no philosophical opinion, which is actually held by anybody whatever, however absurd it may appear and however certainly false, is wholly beneath notice. The mere fact that it is held — that somebody is sincerely convinced of its truth — seems to me to entitle it to some consideration. There is probably, in all such cases, at least some difficulty about the matter, or else nobody would hold the opinion.[1]

Thus justifying myself, I shall proceed. Readers who do not feel the need to know more about this may of course pass on to the next chapter.

The river-of-time metaphor symbolises the flow of time as the steady movement of a current along with all the flotsam it carries

— and the events we experience are depicted by the particular pieces of flotsam being carried along past us by this current. Unfortunately, the movement of such a current can be understood only as a process which itself takes time to happen in; a certain movement of a piece of flotsam which has moved from a particular position to a new position somewhat further downstream can be understood only as a change of position plotted against time. That is, the supposed movement of the current and the flotsam it bears can only be understood *as* movement if it is given time in which to take place. The flow of the river of time cannot be conceived at all unless we introduce a second-order time against which the flow can be measured. If we introduce a second-order time, to be consistent we must picture this new time on the model of a flowing river, and if we do this, the same problem arises in that the movement of flotsam down this second river can be conceived only if we supply still another time in which it occurs. And of course, maintaining consistency dictates that this third time be interpreted as the flow of a river. Clearly the regress is vicious, and the only way to avoid it is to give up the belief that time flows like a river.

The situation is similarly hopeless if we imagine instead the flow of time on the model of ourselves floating down the river past scenes on the banks which represent the events we experience. If we really want to think of ourselves moving down the river of time in this way, we have no option but to think of this movement in terms of change of position plotted against time. Again, the flow of the river must be measurable against a second-order time. And again the regress is generated.

If temporal movement really occurs it would be possible to answer the questions 'How quickly do events flow past me?' or 'How quickly am I passing through time?' This is because movement is a change of position (albeit in this case a change of position in time) and such change, to be change at all, must be supposed to occur at some rate or other. Since the river of time represents lapse of time and not extension in space, the only possible answer seems to be 'Events pass by at the rate of one second per second', or 'We are all moving into the future at the rate of one second per second'. Such answers are incomprehensible, and certainly do not express a rate of change. This is seen all the more clearly if we think of an object moving in space, and someone claims that it is moving at the rate of one foot per foot.

What has gone wrong with the river simile is this: the concept of movement already has embedded in it the concept of time. What I mean by that is that the idea of motion can be understood only if it is appreciated that a change in position occurs during some determinate period of time. That means that the image of movement necessarily cannot picture what we mean by 'time passing', since we have already to know what 'time passing' means to understand what movement is.

There are other related objections to the passage of time being modelled on the river image. If events really are flowing down the river of time they must be doing this at some particular rate. This could only be the case if they were flowing relative to something against which we can stipulate what the rate happens to be. But relative to what are we to suppose the events of history flow? Real rivers flow relative to their channels and banks, but we would be abusing the metaphor to suppose that the channel the river of time flows through represents anything. Someone like Schlesinger may say that the events of history flow relative to the present moment, and this claim solicits my final objection to the river model. In trying to see the passing of events as a genuine flow we immediately face the difficulty already sketched — how can we state what the rate of flow is? And if there really is a *rate* of flow (and if we are not to suppose this, then the idea of real flow would have been abandoned) we are entitled to enquire as to whether this rate can slacken or increase. This is the final objection. Whatever the rate is, it is surely logically possible that it should be either less or more. The idea that we might suddenly begin to move towards the future at twice the rate (say) we are now moving I say is absurd. Could this be clearer? If the normal rate of advance into the future is one second per second, should we then say, if this rate doubled, that we were advancing at the new rate of *two* seconds per second?

These considerations entitle us to reject the notion of temporal movement. They do not entitle us to reject the idea that events change with respect to their being past, present or future: if they do so change, they do not do it by moving out of the future, into the present, and on into the past.

Before moving my discussion on to the tensed view of time, I want to criticise Schlesinger's thought about time — in particular what he says about 'the NOW'.

Note

1. G. E. Moore, *Some Main Problems of Philosophy*, p. 203.

5
Schlesinger's Conception of 'the NOW' Queried

Schlesinger offers an outline of the transient view of time, what he calls the 'common-sense' view, according to which

> temporal points from the future, together with events that occur at those points, keep approaching the NOW and, after momentarily coinciding with it, keep on receding into the past. The NOW is not conceived as some sort of object, but rather as a point in time which any temporarily extended individual experiences as being in the present and which has, so to speak, the spotlight of time focused momentarily on it.[1]

(I do not understand why Schlesinger capitalises the term 'now'. We do not use the expression 'the now' in ordinary language, so it is not in the least clear what the capitalisation is supposed to signify. Schlesinger does not appear to want 'the NOW' to be a technical expression which means something special to philosophers or logicians; as far as I can determine Schlesinger's intentions, 'the NOW' is meant to mean nothing more nor less than the ordinary-language expression 'the present moment'.)

In saying that temporal points and events approach the NOW, Schlesinger is attributing motion to these points and events. If something moves, it needs time in which to move and something against which its movement is relative, as we have just seen in the previous chapter. To provide answers to the questions 'How fast do these events move?' and 'Relative to what do they move?' there is no alternative to postulating an infinite number of times. This is best briefly reiterated. The only way that the alleged movement of events from the future to the present to the past can be understood is by introducing a second-order time by which their movement can be measured. Unless this movement can be measured, it makes no sense to talk of events and times momentarily coinciding with the NOW before moving on into the past. Thus, in the first-order time, we can look at one event, E, and note that first E is future, then E is present, and then E is past. If E is indeed first to be there in time (future) and after that somewhere else (present) we need a second-order time by which to time this change of temporal position. This means that if we look at the events which constitute the second-order time, we will see the event of E's being such-and-such future, the event of E's becoming present, and the event of E's being such-and-such past. If we are to conceive of this second-order series of events as a *temporal* sequence, the events which comprise it must also undergo a similar temporal movement from the future into the past. We have no option but to introduce a third-order time series by which to time the movement of events along the second-order time series. Plainly, a similar need will arise to time the movement of the events which constitute the third-order time series, and we will need a fourth-order time series. And so on. To think that events change in respect to being past, present and future in the sense that temporal points and events that occur at those points keep approaching the NOW, is mistaken.

It is worth noting another criticism of Schlesinger's formulation. He takes 'the NOW' to be a point in time, namely, that point in time which any temporally extended individual experiences as being in the present. If 'the NOW' is indeed such a point in time, can we not ask *which* point in time it is? This can be attempted by applying Schlesinger's own definition of 'the NOW'. If we characterise points in time by their dates,[2] then whichever date we care to choose, we would find that any 'temporally extended' individual at that date would experience it as the present moment, as 'the NOW'. All dates exhibit this feature, for in principle there is an individual alive at each and every date. We could imagine someone who keeps a bizarre sort of journal. In the left-hand column he writes the current date, in the next column he writes the question 'Does the date referred to in the left-hand column denote the present moment?' and in the right-hand column he writes the answer to the question. It is plain that the list of answers written down over the days and weeks, would contain only the answer Yes. There is no date we could find at which someone would say, 'No, this is not "the NOW".'

Schlesinger's 'spotlight of time' is therefore focused on every moment of time, in which case we cannot use the criterion of 'having the spotlight of time focused on it' to distinguish one date from any other as 'the NOW'. The criterion by which that point of time which is 'the NOW' is to be identified in fact picks out each and every moment of time, for at each and every moment of time there is in principle an individual who maintains that he is in the present. The very idea of there being a 'NOW' appears mistaken.

It might be suggested that there is really no dispute between what I wish to hold and what Schlesinger maintains. It might be said that what I said above is addressed to a mis-reading of Schlesinger's remark: he did not mean that the term 'NOW' *names* a particular time — indeed, as we have seen, the idea that 'the NOW' picks out such a particular time cannot be right. The correct reading is rendered by the paraphrase 'The NOW is *any* point in time which any individual experiences as being in the present.'

If that is what Schlesinger meant to say, then we do not disagree. Were we to ask someone 'Is it now *now*?', they would surely answer 'Yes'. What if we asked for instance 'Do all your experiences occur in the present?' Again, the answer has got to be 'Yes'. Whatever we are doing or experiencing, it is always now. Whatever date it happens to be, that date is the present moment. But isn't Schlesinger asserting something more than the trivial claim that whatever is experienced is experienced by the subject in the present? For not *every* point in time is now experienced as being in the present, just *this* point is. And *this* point in time, which is being experienced is thereby different from all other points in time. *This* point in time is the NOW, and all other points are not. 'All moments in time are by no means equal; there is always a privileged moment.'[3] Surely Schlesinger does believe that 'NOW' picks out *a* particular, special, point in time — that point which is being experienced, as opposed to all those other points (noon yesterday, New Year's Day 1987, the three thousandth day from now) which are not.

But as we have already seen, every point in time is special, in that whichever point we care to discuss, any individual at that point would assent to it being 'the NOW'. Schlesinger might reply by asking on 1 May 1987, 'Is 1 May 1987 present?' to which we would reply 'Yes', because it is. Then he would say 'The fact that you assented to *that* date being present, and not to any other which I might have substituted in its place, is what makes it special.'

But that is what we said! *Whichever* point of time is discussed (even 1 May 1987), any individual at that point would agree to its being present. The fact that at any one time we assent to that time's being present, for Schlesinger, gives that particular time the special status of NOWness. All we can say is that if any one time has it, than all times have it.[4]

Notes

1. *Metaphysics*, p. 101.
2. A date is simply a number which states how many time-units lie between the dated event and some previous event relative to which all subsequent or prior events are dated. This is not supposed to establish a technical use for 'date' — that is how dates are actually applied in ordinary discourse.
3. Schlesinger, *Aspects of Time*, p. 33.
4. I look more closely at the expression 'now' in the context of my belief that both space and time are tenseless in Chapter 14.

6

Talking about Temporal Relations

This is the appropriate place to introduce and explain the terminology that will prove essential for developing my discussion.

There are facts about the temporal relations between events expressed in terms of precedence and subsequence, or expressed with the phrases 'earlier than' and 'later than'. Given a whole list of temporal facts about a number of events we would be able to assemble those events in the correct temporal order, starting with the event which precedes all the others and ending with the one that occurs later than all the others. To take a simple example of how we would construct such a temporal sequence, suppose we were given a batch of cards representing distinct events, labelled E_1, E_2 . . . E_{20}. (There is no reason why we should not think of these cards as photographs: E_1 might show a man getting out of bed, E_2 the same man eating a meal, E_3 someone else opening a letter, etc.) We are also in receipt of a list of temporal facts — 'E_1 is earlier than E_{19}'. 'E_{20} is earlier than E_5', and so on. I think it is fairly clear that if we were given enough facts we could lay out the cards starting with the one that represents the earliest event and ending with the one that represents the latest event. There on the floor would be a representation of a temporal sequence of events, ordered by the relation 'earlier than'. Notice that we cannot tell which card represents the event that is taking place in the present. *Our* relation to these events (supposing the cards to represent real events) is completely indeterminate given just our list of temporal facts. Indeed, the card representing the most recent event (that is, the one that occurs later than all the others) might represent an event that happened centuries ago.

Following McTaggart and later writers, I shall call a sequence of temporal events generated in this way, a *B-series*, and the relation between events in the B-series I shall call *B-relations*. Events between which B-relations hold can be said to be *B-related*.

Suppose now that we are given the same set of cards, E_1 to E_{20}, but a list expressing their temporal relations by stating the degree of each event's pastness or futurity. The list might run 'E_1 is five days past', 'E_{19} is two days past', 'E_5 is present', 'E_{20} is three days future' and so on. Once again, it seems plain that (given enough of these facts) we can order the cards correctly. The sequence is not ordered by a single relation, but by each member of the sequence having a unique relation to the present moment (two days past, three days future, or what have you). A sequence of events generated in this fashion I shall term an *A-series*.

An event is ascribed a position in the A-series[1] by establishing a temporal distance between the present and the time when the event occurs. 'E occurred last week' establishes just such a distance, and such an ascription of position can be more or less vague or accurate. If it is true that E occurred last week, it might also be true that E occurred exactly five days ago, and that E occurred last Wednesday afternoon. Thus we can define a technical use for the phrase 'tensed expression': such an expression ascribes a position in the A-series to a particular event, and that event's particular position will be referred to as its *A-determination*. Thus 'The Black Death occurred centuries ago', 'My wedding took place last year' are tensed expressions. This technical use appears to incorporate the familiar grammar-lesson use of 'tensed expression', which refers to sentences like 'He *fell* off his horse', 'He *will* pay up', where certain events (falls from horses, the honouring of debts) are given a temporal position relative to the present. But grammar-lesson tensed expressions like 'If only I had been good, he would have been kind' appear to be excluded, since an expression like that is not aiming to place an event on the A-series (its aim is to do something different). My interest will be in tensed expressions which place events upon the A-series. These expressions work by naming parts of the A-series ('last week', 'the future', 'yesterday', 'today') or by doling out as it were chunks of time which take us either way from the present a certain temporal distance ('three days ago', 'one week from now', 'three thousand years ago').

Another way of locating events on the A-series is to give their dates. In 1987, we could say truthfully that the first man on the moon landed there eighteen years ago. We could also say that he landed there in 1969. To give an event a date does not at the same time give it a position on the A-series; giving the date of an event as 1969, say, tells us nothing about its pastness or futurity. To know about that, we need also to know the date of the present moment. Suppose we know that it has been said 'E occurs in 1969'; since we know that the present date is 1987, we can properly conclude that E is past — eighteen years past, to be precise.

I will follow custom and call the sorts of tensed expressions I am interested in *A-statements*, and sometimes I will simply call them *tensed statements*.[2] Thus an A-statement, such as 'E is past', or 'E is future', ascribes to the event it mentions a certain location, more or less vague, on the A-series, either 'downstream' or 'upstream' from our current position. *B-statements* then, are statements which ascribe to particular events positions on the B-series in virtue of the relation expressed between the mentioned event and some other event, again, more or less vague. 'E_1 is earlier than E_2' is therefore a B-statement, even though E_1's whereabouts on the B-series could be anywhere amongst all those events of which it would be true to say that each occurs before E_2. The ascription is vague. 'E_1 is three days earlier than E_2' is also a B-statement, but one which is more precise in its ascription.

The death of William the Conqueror is earlier than the death of Charles I. Like all B-statements, if this statement is true (and it is), it is true no matter when uttered. There are no times at which someone could use this statement to assert something true, and other times when it could be used to assert something that is false. We can express this feature of B-statements by saying that if a B-statement says something true, it is true 'eternally' or 'timelessly'.

Since the B-series is nothing more than a sequence ordered by the relation 'earlier than' it does not make sense to suggest that a description of part of it, for example, that E_1 is earlier than E_2, could change its truth-value. To think that it could would be to think that the events on the B-series could somehow shuffle round and change their relations. And if someone sees that as a possibility he would have to hold as well the possibility that even though E_1 was experienced to occur before E_2 it will not always be true to say that E_1 occurs before E_2, and that seems patently absurd.[4] Events are fixed in their relations to all other events. If someone says that B-statements are 'eternally' true, that, at the very least, must be what he means. If people ever dispute the ordering of events, that is not because the events involved are somehow shifting about — there is no dispute about the fact that the events are fixed. The dispute is about which description correctly applies.

A-statements, unlike B-statements, *do* have different truth-values at different times. It is now 1987, and if I said 'The Great Fire of London (of 1666) occurred 321 years ago' what I said would be true because the Great Fire of London did indeed occur 321 years ago. But if I had asserted this last year, in 1986, what I said would have been false. The A-statement 'The Great Fire of London occurred 321 years ago' is true throughout 1987, but false at all other times. This will be found to be true of all A-statements; there will be only certain times during which it would be possible to assert the A-statement and in one and the same breath say something true. Admittedly some certain times might be fairly enormous; for instance, it is true now that the Great Fire of London happened in the past, and it will be true for the rest of eternity, but before it occurred it would have been false to maintain that it occurred in the past. In contrast, the B-statement 'The Great Fire of London is earlier than the Second World War' would be true no matter when stated.

Notes

1. What Gale calls an *A-determination*; see Gale (ed.), *The Philosophy of Time*, p. 67; or what D. H. Mellor, *Real Time*, calls a *tense*. (Thus 'last week', 'tomorrow', 'next year' are for Mellor tenses, cf. p. 16, and for Gale A-determinations.)

2. I will of course distinguish temporally tensed statements from spatially tensed statements.

3. We should note that events can also stand in the B-relation 'simultaneous with'. This relation can be expressed in terms of 'earlier than' because we can say that E_1's being simultaneous with E_2 entails both E_1 and E_2 being earlier than some third event, E_3, to exactly the same degree.

4. cf. Richard M. Gale, *The Language of Time*, p. 9, 'The death of Plato, for example, cannot through diligence and hard work sneak up on the death of Queen Anne, for if one event is earlier than some other event by so many time-units it is always the case that the one is so many time-units earlier than the other.'

7
The Tensed View of Time

Time is not dynamic in the sense that events and the dates at which they occur really move through time, progressing out of the future to coincide with the present, then moving on into the past. This leaves the second sense of 'dynamic' to investigate: this is the idea that events and dates are dynamic in the sense that they change with respect to being past, present and future. Those who believe that events and dates really do this I refer to as holding the tensed view of time.

On this view, phrases such as 'the flow of events', 'the passing of time' are taken not to refer to real movement, logically analogous to the movement of objects in space (such as logs floating down a river), but are taken to refer to the fact that events have the non-relational properties of 'pastness', 'presentness', and 'futurity'. Not all events are perceived to be occurring in the present moment. Those that are, which are indeed said to be present, are supposed on this view to have the property of presentness. Those that are not, are either past or future, and what makes them so is their possessing either the property of 'pastness' or 'futurity'. And if we conceive of events as lying on an A-series, it follows on this view that the A-determination that an event has is determined by whichever property (pastness, presentness, futurity) it happens to have. If event E lies on the A-series in that portion designated by 'yesterday', it has that A-determination because it has the property of pastness.

This view of time and events is properly called 'dynamic' since, although no real movement is attributed to events, events are thought of as constantly changing their A-determinations (as a consequence of changing their temporal properties), in contrast to the static view, on which all events retain their particular B-relations with all other events for the whole of eternity. There are no occasions upon which an event could be less early, say, than some specified event than it was on another occasion. However, conceived *dynamically*, if at one time an event has a certain A-determination, it will not have *that* A-determination (but another one) at any other time. In merely thinking of events as lying on an A-series, we do not seem to be committing ourselves to a belief in either the tensed view of time as outlined above, or the transient view as discussed in earlier sections. The transient view is logically incoherent. But what of the tensed view?

Is it right to think that events have these strange properties of pastness, presentness, and futurity? And there is also the question as to whether ordinary physical objects might have the very same properties. For instance, is it true that the original St Paul's Cathedral, destroyed in 1666, now has the property of pastness? And does a building which is to be erected next year now have the property of futurity?

An attack upon the view that events have the properties of pastness, presentness, and futurity, is an attack upon the very substance of the dynamic view of time. The dynamic theorist must mean something more than 'there are A-determinations', if by 'A-determination' he means that people refer to events as having a certain temporal distance from the present moment. An event, E, is ascribed an A-determination when we say 'E occurred three days ago.' The fact that such expressions are used does not in itself commit us to a particular metaphysical view of time. Someone who holds the static view of time would say that 'E occurred three days ago' is not made true (supposing it is true) by E's having a special property (of pastness) but by the fact that E and the occurrence of someone's saying 'E occurred three days ago' stand in a particular B-relation — that is, E is earlier than the utterance, by three days. If that is all the dynamic theorist thinks an event's having an A-determination comes to, then there is no real difference between what he holds and what a static theorist holds. For there to be a difference, the dynamic theorist seems committed to the view that there is something about the events themselves that gives rise to their having the particular A-determinations they in fact have. What this difference might be, we have already seen, cannot be explained in terms of 'the NOW', or of events moving. The dynamic theorist is reduced to defending the tensed view of time, on which statements such as 'E is in the past' or 'E is in the future' are cashed in terms of E undergoing changes with respect to the properties of pastness, presentness, and futurity; that is, the change in truth-value from true to false of the statement 'E is in the future' as E becomes present is dependent upon E's shedding futurity and gaining presentness. The truth-value changes because there is a change in something else — that is, a change in which temporal property E as a matter of fact possesses.

I shall later have occasion to use the expression *tensed fact*. Someone who believes that a particular event is either past, present or future I shall describe as believing that the event is *tensed*, and this alleged fact, of the event's being either past, present or future, I shall call a *tensed fact*. Thus an A-statement, which can also be called a (temporally) tensed statement, is supposed on the tensed view of time to express a (temporally) tensed fact.

8

Objects, Events and Properties

It is instructive to pause and ask why anyone should feel like attributing properties to *events*. There is a little more that can be said over and above the rather obvious point that it is tempting to model statements like 'My wedding is past' on statements like 'My dog is brown'.[1]

An object changes when it has a certain set of properties at a certain time, and a different set at a later time. 'Object' here needs to be taken in a broad sense. We often refer to changes in physical objects, such as apples, less often stars, and still less often objects which occupy the hazy borderline of the category, such as electrons or alpha particles. The other category of 'object' that we refer to as sometimes changing is that of mental entities, including pains. A pain may change from being a constant one to being a throbbing one; and one's mood may change from well-being to depression. Objects, then, taken in this broad sense, change in so far as they can gain and shed properties. An object gaining and/or shedding properties constitutes an event. An event then (speaking rather imprecisely, but precisely enough for present purposes I think) is a change in some object.

What the tensed theorist wants us to believe is that events, which are changes, can themselves change; namely, they can change their temporal properties of pastness, presentness, and futurity. The idea of a change changing may strike one as fishy. Smart has said, '. . . events are happenings to things. Thus the traffic light changed from green to amber and then it changed from amber to red. Here are two happenings, and these happenings are changes in the state of the traffic light.[2] That is, *things* change, *events* happen. The traffic light changes, but the changing of the traffic light cannot be said to change. To say that it does or does not change is to utter nonsense.'[3]

A tensed theorist who believes that events possess temporal properties will not be moved by Smart's comment. He will happily admit that whereas it is correct to speak of the traffic light changing its colour, it is a mistake to speak of that change itself changing colour. But that is no argument against the view that events can change with respect to properties which they do have. An event cannot change its colour because events are not in that class of particulars which are coloured. But events are capable of changing their temporal properties. Just because events cannot change with respect to colour, it does not follow that they cannot change with respect to the temporal properties of pastness, presentness, and futurity.[4]

The urge to see events in this way, I suggest, arises from our common everyday experience of giving and receiving what can be called 'explanations-via-properties'. Some properties we point to are non-relational. Thus, when asked 'Why is the bookshelf bowed?' we may answer 'Because the books are heavy.' Our explanation as to how it is that the shelf comes to be bowed succeeds by pointing to the heaviness of the books. Given our understanding of the behaviour of physical objects and various materials, this explanation is taken as an adequate answer to the original question. Psychological states can also feature in similar explanations. Thus if S is asked why he is so irritable, one acceptable explanation is that the pain he has had now has the property of being more intense, or more stabbing; we are pointing to a property of his pain. Other instances of 'explaining-via-properties' work by pointing to rational properties. For instance, the reason why S is playing in such and such a position in the basketball line-up is that S is taller than all the other players in the team.

If someone is thinking about time, and he looks for an explanation as to why a past event (the Battle of Hastings, for instance) or a past object (Cleopatra's barge, for instance) are not presently perceivable, he might conclude that a satisfactory answer is to say that the event or the object is *past* — meaning that its having the property of pastness accounts for its present unavailability for scrutiny by the eyes. Such an explanation, which seemingly succeeds by atrributing the property of pastness to the event in question, is logically similar to many possible explanations which work by ascribing properties. It is true to say that the property of moving at great speed possessed by the bullet accounts for the bullet's unavailability for scrutiny by the eyes. 'The bullet cannot be seen because . . . .' And the explanation is completed by attributing the property of speed to the bullet. It seems on the face of it plausible that the explanation 'The past event cannot now be witnessed because . . .' ought to be completed in similar logical style.

My point is to show how natural it is to think that events can have temporal properties. This is because so many of the explanations we make and hear succeed by ascribing properties. That's why it's easy to believe that explanations about non-present objects and events ought to work in the same way.

The belief that they do, is all the more fostered by the fact that in some contexts we seem happy to attribute non-temporal properties to events. We say such things as 'That was a happy occasion' and 'The journey was boring.' It looks as though we are ascribing properties to the events themselves — we seem to be labelling the occasion as happy, and the journey as boring, in exactly the same way that we labelled the books as heavy, the basketball player as tallest in the team, and the pain as intense. But we could have said, 'That occasion made me happy' and 'The journey bored me.' These expressions assert that it is the subject of the experiences who has the properties of happiness or boredom. These are the better expressions. 'The journey was boring' is true only in so far as *someone* happens to be bored by the journey. The boredom is in the traveller and not the journey itself. The view that the journey can in *itself* be boring fails when we notice that it is always possible that at least one person on the journey finds it interesting.

Having so far given just a perspective on why it is easy to think that events have temporal properties, I have said nothing that counts against the view that they do possess these properties. This will now be attempted.

Notes

1. cf. Broad, 'Ostensible Temporality', p. 127f. and Flew, 'The Sources of Serialism'.
2. [i.e., at one time certain properties are attributable to the traffic light, and at later times, different properties are attributable.].
3. J. J. C. Smart, 'The River of Time', p. 216.
4. cf. Schlesinger, *Metaphysics*, p. 103.

9
The Unreality of Temporal Properties

There are necessarily no such things as temporal properties. It is a mistake to think that the terms 'pastness', 'presentness' and 'futurity' refer to properties that events can have, in the same way that 'redness' refers to a property that apples (and other things) can have. There are a number of things I will say to expose this mistake. And having exposed it, as explained in Chapter 7, the tensed view of time will be refuted. By default, this will leave the static view as being the correct view; whether or not the static view says everything that we think an adequate theory of time should say cannot be demonstrated by showing only that the dynamic view is wrong. More will need to be said, since without careful thought, it will not be clear what the implications will be for tensed statements (statements which ascribe to events A-determinations, such as 'E occurred three days ago'). It might be thought that if the dynamic view, which gives primacy to tensed expressions, is in fact wrong, then tensed statements must be translatable by tenseless statements (statements which do *not* ascribe A-determinations to events, but which express the B-relations that obtain between them, as in 'E_1 is three days earlier than E_2'). Gale thought this,[1] and since he shows, correctly, that tensed statements cannot be translated by tenseless ones, he believes, wrongly, that the tensed view of time can be defended. I will look at this question of whether tensed statements can be translated into tenseless ones in Chapter 13.

How can it best be shown that there are not, and could not be, temporal properties? Someone who believes that an event can have the temporal properties of pastness, presentness and futurity, has fallen into a confusion created by the same surface grammar of sentences such as 'That apple is red' and 'That event is past' and others like them. The apparent parallel between 'That apple is red' and 'That event is past' is maintained in saying for instance 'The apple's being red is why I see a red apple', and 'The event's being past is why I have a memory of it.' The similar surface grammar of the statements leads some to see pastness, presentness and futurity as *properties* of events just as much as redness is a property of some objects. If 'That apple is red' is allowed to take the lead, as it were, we might be inclined to assimilate 'pastness' with 'redness' and other familiar properties.

It is possible to show the error of making that assimilation by saying a number of things which I think demonstrate that the depth grammar of 'pastness' is very different from the depth grammar of those terms which we accept denote properties. I will show that this is so later in this section. The consequence for the tensed view of time is that although the statements 'E is occurring now' said today, and 'E occurred yesterday' said tomorrow, are true, what makes them true is not E's shedding the property of presentness and gaining the property of pastness.

Firstly, let us imagine two physical objects, A and B, which are located such that A is to the west of B, by ten yards, say, or what comes to the same thing, B is to the east of A, by ten yards. If this state of affairs is given, we would of course be making a true statement if we said 'A is to the west of B, by ten yards.' We would not say that what makes that statement true, apart from the spatial positioning of A and B, is some extra fact such as A's possessing the property of 'westness' and B's possessing the property of 'eastness'. If we correctly say that A is ten yards west of B, we would not want to be understood as attributing the strange spatial properties of 'westness' and 'eastness' to A and B, or be referring to some further fact beyond A and B being thus spatially related.

An analogous argument can be addressed to the situation where E happens to be three days past (say). That being the state of affairs that obtains, of course we would be making a true statement if we said 'E is three days past', or, a little more colloquially but logically equivalent, 'E occurred three days ago'. We would not say that what makes that statement true, apart from the temporal positioning of E and the event of our statement about E, is some extra feature of the situation such as E's possessing pastness and the event of our statement possessing presentness. If we correctly say that E occurred three days ago, we would not want to be understood as attributing the temporal properties of pastness to E and presentness to our statement about E, or as referring to some other fact.

This, unfortunately, will not convince a tensed theorist of his error. He agrees that there are no spatial properties such as 'westness', but the analogous remark about events in time would leave him unmoved. His position is that what makes 'E occurred three days ago' true is indeed the fact that E is past, that it is correct to predicate pastness of E. Our argument simply denies what the tensed theorist asserts, and just to make the denial would not in itself compel him to change his view. It would be better if we could show that the belief in temporal properties, the belief that pastness, presentness and futurity can be predicated of events, is incoherent.[2] This I believe can be done.

The supposed temporal predicates of 'is past', 'is present', 'is future' (by means of which the tensed theorist attributes temporal properties to events) are incompatible; there is no event of which we could at this moment correctly predicate 'is past', 'is present', 'is future'.

This remark would be taken by the tensed theorist as a piece of facetious silliness. Of course there is no event to which we could apply all three predicates — not all at once. Temporal predicates work like any other predicate which attributes a state or a property to a particular. 'John is ill' is true only in so far as there is a time at which John is ill. In ordinary language, if we make a statement of the form '*a* is F', what we mean is '*a* is F *now*', that *a* being F and our saying that it is occur concurrently. (Of course we can make true statements about particulars at times other than when the particular is in the condition we mean to attribute to it, by saying, for instance, 'John was ill in 1982.') The tensed theorist would explain that when he says we can predicate pastness, presentness and futurity of an event, he does not want to be taken to be saying that an event can be past, present and future all at the same time. That would, indeed, be a very silly idea. But just as 'John is well' is true at one time, and 'John is ill' is true at another time, an event, E, can be future at one time, present at another, and past at a third time. There is now no contradiction in the claim that events can be past, present and future at different times.

Let us try to accommodate this view. If E is past, present or future at different times, we can ask when those times actually are. Since all times are either past, present or future, *those* times[3] must be either past, present or future. Thus, E is past in the past, present or future, or E is present in the past, present or future, or E is future in the past, present or future. Altogether we can predicate nine second-level temporal predicates of E: E is past in the past, E is past in the present, E is past in the future, and so on. We found that the first-level predicates 'is past', 'is present', 'is future' are incompatible in that if E is to have them, it must have them at different times. But the same is true of the second-level predicates — E can have them only if it does not have them all at one time; E cannot both be future in the present, and past in the present. Once again we can ask for the times at which E has these second-level predicates, and *these* times must themselves be either past, present or future. Thus twenty-seven possibilities are obtained: E is past in the past, in either the past, present or future; E is past in the present, in either the past, present or future, and so on. So at what times can we predicate these third-level predicates? In attempting to provide an answer, the tensed theorist is forced to predicate fourth-level predicates of E. In explaining how E can have the incompatible temporal properties of pastness, presentness and futurity, he has commenced on a regress which at no stage can count as an explanation, and which at every stage we can press for an answer to the question 'And *when* (past, present or future) does E have *that* temporal predicate (whatever level of predicate this might be)?' The only answer the tensed theorist can offer itself requires an answer to the very question it was supposed to settle. What is responsible for this state of affairs is the original insistence that 'is past', 'is present', 'is future' predicate temporal properties of events.

It might be objected that the 'is' in 'E is past in the present' (for example), is not the 'is' of predication, that what I have called a second-level predicate is not really a predicate. If it is not, it is not possible to ask at which times it is correct to predicate it. In which case no regress results, because it would not be necessary to offer the third-level 'predicates' as the answer to 'When (past, present or future) is it true that E is /past, present or future/ in the /past, present or future/?' Certainly, 'E is past in the future' (for instance) — which is logically equivalent to the more colloquially agreeable 'E will appear to be past from times in the future' — is an unusual sentence, and it is not clear that in using it there is any implication that a 'two-place property' (whatever that is)[4] is being predicated. However, since the tensed theorist would agree that 'is' in 'John is ill' and in 'John is ill in the past' is the 'is' of predication, he would have to accept that the 'is' in what he wishes to argue are logically analogous statements, 'E is present', 'E is present in the past'[5] is likewise the 'is' of predication. The objection fails.

What we can conclude, then, is that the insistence on the part of the tensed theorist that 'is past', 'is present', 'is future' predicate temporal properties of events is a mistake. This is so because we cannot make sense of the claim that a particular event can have those properties predicated of it at different times. In which case it is not obviously clear as to how we should understand the use of tensed statements: if they do not attribute temporal properties to events, what do they do? If they are not made true, when they are true, by tensed *facts* (by events being *in* the past, say, that is *having* 'pastness') then what does make them true? This question will be discussed in Chapter 12.

Schlesinger acknowledges the above argument,[6] but is not convinced by it. The apparent regress, Schlesinger would argue, arises because we attempt to apply the incompatible temporal predicates to one event all at the same time. If we resist doing such a silly thing, no problem results, and the regress never begins. '. . . surely there is no reason to believe that "past" and "future" as such can be assigned to the same event simultaneously.'[7] But as I tried to show, the regress results because of our attempts to establish at which times we should attribute the first-level temporal predicates. What could be clearer than the truth that if E is either past, present or future it must be past, present or future at a time which is either past, present or future?

A second line of defence is tentatively offered but not vigorously argued for. Schlesinger suggests that for a regress to be philosophically disturbing in the way I require, it has to be one such that 'we can raise a new problem after every solution rather than that for every problem there is a solution . . .'[8] This idea cannot be applied in the present case, because as I see it, the regress we have is not a string of problems and solutions, but just a string of problems. The problem with the set of three first-level temporal predicates is that they cannot apply to an event at just one time. The set of nine second-level predicates (E is /past, present or future/ in the /past, present or future/) offers no solution because these predicates are subject to the same difficulty in that they also cannot apply to an event at just one time.

As promised earlier, I will conclude this section by making some remarks which illustrate the fact that 'pastness'[9] has a very different depth grammar from those terms which we accept denote properties. What I mean by that is that the assumptions it would be right to hold and the questions it would be appropriate to ask with respect to apples and their redness, for instance, do not hold good for events and their pastness. The following comments, much less formal than the argument rendered in the first part of this chapter, point to the fact that the notion of temporal properties is incoherent, the finding which the first argument was meant to establish. Even though I feel convinced by the first argument, others might be less enthusiastic. In so far as I think other reasons, about to be given, can be held to support the conclusion that the first argument aimed to establish, any flagging enthusiasm ought to be restored, if not in the first argument then in its conclusion. In the same way that if two or more experiments support the same theory we will feel happier about embracing the theory, if two lines of argument support the same philosophical conclusion, wc ought to feel happier about accepting that conclusion than we would offered with just one of the arguments.

(1) Ignoring those cases where someone receives information from someone else, written or spoken, about the situation, when someone says that a certain physical object, A, is red (say) it is right to assume either that the object is in his presence and that he is seeing it, or that it *was* in his presence and that he *has* seen it, and that A is the sort of object which, having been red, is still likely to be red. His saying that A is red is dependent upon his having, or having had, certain sense experiences.

It is usually possible to verify that A is red, because in usual circumstances the fact that A is red is publicly observable. We perceive directly what the case is. It would be right to assume that someone's seeing that A is red might be prevented by usual contingencies such as its being too dark, another object obscuring the view, A's being somehow concealed, etc. Appropriate questions obviously attach to such assumptions, such as 'Is it too dark to see?' etc.

But when it comes to a certain event, E, being past, the sort of assumptions and queries that apply to the case of A's being red have no bearing on the situation. If the fact of E's pastness is publicly observable, it is so in a very different sense from that of A's redness being publicly observable. We cannot stare at E and see its pastness. One simply does not *see* that E is past in the way that one does *see* that A is red, even though we do talk of 'seeing' that E is past. But 'seeing' here has the sense of 'understanding' or 'knowing'. We do not appreciate E's pastness through the having of a sense experience. Of course, we may see it written down that

E is past, or hear it said. But it is not E's pastness which is affecting our eyes or our ears, whereas it is A's redness which affects our eyes. There is nothing we can do, nowhere we can go, to get E's pastness to affect us in any way.

It might be said that we can see E is past in the sense of having memory experiences. But we need to point out that 'seeing' here is not *seeing*. One turns the 'eye of memory' upon a past event in only the highest of metaphorical senses. The image that is 'seen'[10] in a memory experience is not an image of E's pastness, and neither is it a publicly observable entity. 'Seeing' that E is past has only and exactly the sense of 'understand' or 'know'. 'Seeing' should not be assimilated with *seeing*. The procedures for deciding that E is past are altogether different from the procedures for deciding that A is red.[11]

(2) It is correct to assume that if a physical object has a certain property, if A is red, we have no difficulty in imagining any number of occurrences that would result in A's no longer being red. If E is past, or if B (a past object, or a past 'temporal slice' of an object) is past, there is nothing we could conceivably do to make E or B either present or future. The temporal properties of events and objects are logically beyond our power to tamper with, whereas, in principle there is always something that can be done to change the properties in a physical object.

This counts against the view that 'pastness', 'presentness' and 'futurity' denote temporal *properties*. The word 'property' when used on its home ground refers to states of physical objects that 'appear' to and 'disappear' from our senses depending upon the influence of natural phenomena or an agent's actions. So-called temporal properties are not in that category.

(3) If something has properties, then it must exist. If someone affirms correctly that A is red, there must actually exist this A which is red.[12] The view that there is a parity with respect to E and its pastness seems very unsatisfactory. Since all events have one of the three temporal properties (pastness, presentness, futurity), it follows that all events exist. Yet anything which exists is a possible object of experience.[13] But very few events are possible objects of experience (only presently occurring ones are). Someone who believes that events have temporal properties seems committed to holding also that all events can in principle be experienced. On this view, the fact that very few events are actually experienced (present ones), is a contingent fact; that is, if each event that ever has been or will be is in principle an object of current experience,

if, as a matter of fact, a certain event is not being experienced, that is a mere contingent eventuality.

This cannot be right. To hold that a past event, E, could, as a matter of contingent fact, be the object of a current experience, is to say that E has the property of pastness but might also have the property of presentness. (What is meant by E having the property of presentness is that E can be an object of experience.) If the possible experiencing of E were to become an actuality, then E would have the property of pastness as well as the property of presentness. This is to say that E is past, and in the same breath to say that E is not past — surely a contradiction. Further, this view requires it to be possible that E be a possible object of experience *and* a possible object of memory; if E is being observed, E can also (possibly) be remembered as having been observed.

This outcome of the idea that pastness, presentness, and futurity are properties of events defeats the tensed theorist's purpose in introducing the temporal properties as properties in the first place. His attempt to show that A-statements change their truth-value as the consequence of events gaining and shedding the various temporal properties fails because his view that 'pastness', 'presentness' and 'futurity' are *properties* allows for the possibility that any event can be experienced, remembered and anticipated all at once. The view is incoherent.

(4) If A is red, it would be right to assume that, depending upon what happened to it, it might later be blue, and later still red again. There is no parity with respect to events and their supposed temporal properties. If an event, having been present, is now past, it cannot thereafter acquire the property of presentness again. But why should that be? Well, it is simply how things are. Even people who believe that history repeats itself do not believe that the very same events occur all over again. There remains the need for an explanation for why it should be that an event which possesses the property of pastness cannot ever again possess the property of presentness.

(5) Given that E is now occurring, we can conclude that it will be past (that having the property of 'presentness' it will next have the property of 'pastness'). This is a matter of logical necessity. Any similar occurrence with respect to objects — for instance, that an object having been red (say) will next change to blue — is a matter of contingency.

(6) Two events may both be past without being to the same degree past — one can be more past than the other. Similarly for

future events. How is this represented on the temporal properties view? It seems not to be. Ought it to allow an infinite set of temporal properties — one for each position on the A-series? Or would it be better to say that one event being more past than another is the result of its having *more* 'pastness', of it containing more of that stuff such that when an event has some of it, it is past? Such adjustments would be *ad hoc*, besides which it is not clear that such adjustments are satisfactorily comprehensible.

(7) This final objection to the temporal properties thesis strikes me as particularly forceful. The comment is Smart's, and I quote it from Adolf Grünbaum's paper:

> If past present and future were real properties of events [i.e. properties possessed by physical events independently of being perceived], then it would require [non-trivial] explanation that an event which becomes present [i.e. qualifies as occurring *now*] in 1965 becomes present [now] at that date and not at some other (and this would have to be an explanation over and above the explanation of why an event of this sort occurred in 1965).

All these considerations indicate that the use of the statement 'E is past' cannot provide grounds for 'pastness' being a property, or for events belonging to that category of entity that can possess properties.

It might be objected that I have dwelt too much on the property word 'red' and that no one would want to defend the view that the workings of temporal properties share much in common with the workings of properties like 'red'. This exposes the whole problem from another angle. Unless temporal properties share some fundamental set of features characteristic of other properties, the idea that events can 'acquire' them and 'shed' them becomes meaningless. This is just another way to express what I have been claiming — if the depth grammar is dissimilar to *that* extent, it is incoherent to hold that so-called temporal properties really *are* properties which particulars can 'have'.

Notes

1. Gale has written extensively about time. See for instance his 'Tensed Statements'.

2. I do not mean that 'E is past' for example is really incoherent, or somehow 'wrong'. What is wrong is taking such statements as signs that there are temporal properties.

3. It might be objected that I attribute to the tensed theorist the predicating of temporal properties to *times* themselves. That this can be done on the tensed view follows from the fact that it is plainly absurd to claim that my opening my Christmas presents is future, but that the 25th December is not future, or is future in a different sense.

4. Whatever confusion there might be over this is not relevant to the discussion.

5. That is, 'John is ill' is analogous with 'E is present', and 'John is ill in the past' is analogous with 'E is present in the past'.

6. In his book, *Aspects of Time*. See pp. 47 ff.

7. *Aspects of Time*, p. 49.

8. Ibid. He looks into issues raised by regresses in Chapter 8 of *Metaphysics*.

9. I hope there will be no objection to my concentrating on 'pastness' in these remarks. Since the tensed theorist believes that A-statements change their truth-values as a result of the events referred to gaining and shedding temporal properties, if I can show that 'pastness' ought not to be taken as referring to a temporal property, enough will have been said.

10. If indeed an image is 'seen' (or had) at all.

11. The considerations of the last two pargraphs apply equally well to past objects, as well as to past 'temporal parts' of existing objects.

12. This is not strictly true of fictional entities. Although Sherlock Holmes is intelligent, we should not conclude that Sherlock Holmes exists. Even though we are saying something true when we say that Sherlock Holmes is intelligent, we are not in saying that attributing a property (intelligence) to an existing entity — obviously, since Sherlock Holmes does not exist (part of what we mean by 'fictional'). 'Sherlock Holmes is intelligent' is true because it is an ellipsis of a more complicated statement which offers a more complete picture of what is being claimed, namely, that 'Arthur Conan Doyle wrote stories about a fictional character, Sherlock Holmes, who, in the stories, is intelligent.'

13. Electrons exist, but they are not possible objects of experience. But the existence of electrons is logically required by applying certain theories of science to phenomena (such as active television screens) which are objects of experience. Sadness exists even though what we see is people's 'sadness-behaviour'. The view that sadness is a state of the mind is contentious precisely because we are not sure what should count as evidence for the existence of a state of mind which logically cannot be observed. The view that we can observe evidence from which we can infer that events possess pastness (as distinct from the view that we can have evidence that such and such has occurred) is too unclear to merit discussion.

14. J. J. C. Smart's remark appears in his book *Philosophy and Scientific Realism*, Chapter 7, and is quoted by Grünbaum in his paper, 'The Status of Temporal Becoming', p. 344. The extra bits [in square brackets] are by Grünbaum.

10
Summary

I have argued against the view that 'is past', 'is present', 'is future' can be used to predicate temporal properties (that 'pastness', 'presentness' and 'futurity' refer to properties) in two ways. In the first part of Chapter 9 I tried to show that the attempt to elucidate the claim that a particular event can be past, present and future at different moments of time results in a regress of failed solutions, each aiming to answer the question 'When (past, present or future) does E have *this* temporal predicate (whichever level of predicate along the regress that might be)?' and in so doing exposing itself to the very same question it was supposed to answer. And in the second part of Chapter 9 I made a number of remarks which I think establish the fact that 'pastness' in particular does not function in our language like paradigm property words such as 'red'. In other words, the depth grammar of the so-called temporal properties is completely different from the depth grammar of words like 'red'. Because 'temporal properties' function completely differently from the way usual property words function, the view that events can 'have' temporal properties, 'gain' them and 'lose' them doesn't make any sense, because what we mean by 'having a property', 'gaining/losing a property' has application only in our talk about properties such as 'red' which as a family share essentially the same depth grammar. This indicates that when we say an event is past, we should not be taken to be predicating 'pastness' of the event; if we say that an event *has* 'pastness', its 'having' such is a metaphorical extension of the expression 'to have', usually used to predicate properties of objects and people.

These findings, and others discussed earlier criticising the transient view of time (in Chapters 5 and 6), leave us with a negative thesis which states that events change their A-determinations *not* because they gain and shed temporal properties (the tensed view), and *not* because they really move relative to 'the NOW' (the transient view). In other words, the ever-changing panorama of events which everyone experiences, which is constituted by objects forever changing their properties,[1] is *not* explained by holding that events are forever changing their temporal properties. And neither is it explained by believing that we are rushing towards future events and away from past ones, or that we are stationary and it is the events that do the rushing.

Note

1. These would include relational properties.

11
McTaggart and Change

J. M. E. McTaggart thought that change can occur only if events have A-determinations which they can change. Since he thought that events cannot have A-determinations (so obviously cannot change with respect to them), pursuing a version of the argument I presented in the first part of Chapter 9, he concluded that change is an illusion, as well as that time is unreal. To say that time is unreal, is McTaggart's way of encapsulating the idea that whatever entities there may be, none 'can be temporal'[1] — which is equivalent to the view that there are no temporal relations. McTaggart finishes with the words, 'Nothing is really present, past, or future. Nothing is really earlier or later than anything else or temporally simultaneous with it. Nothing really changes. And nothing is really in time.' I do not agree with McTaggart, because although I hold with him that events do not have A-determinations which they can change, I do not believe that that fact supports any argument for the view that there are no temporal relations, and the argument which McTaggart uses to conclude that there are no temporal relations is valid only because he incorporates a false notion as to what change is. So despite his logic being correct, his conclusion is false.

His argument comes in two sections. In the first, he aims to establish the premiss that if time is real, events must change in a certain way. (This is so in consequence of the undisputed claim — albeit a vague one — that the essence of time is change: '. . . there could be no time if nothing changed'.) In particular, events must change with respect to their A-determinations. Believing that he accomplishes this, McTaggart commences his second section attacking the tensed view of time showing that it is incoherent, and therefore false, that events can have A-determinations in respect to which they change. Having already established in the first section that if time is real events must so change, he can conclude by *modus tollens* that time is unreal. McTaggart's strategy, briefly reiterated, in this. The only possible conception of time is that of the tensed view. Since the only possible conception of time is incoherent, time is unreal. Noting that I think McTaggart's criticism of the tensed view of time is essentially sound (in the second part of his paper), my real interest is to attack what he says in the first section of his discussion.

It was McTaggart who, in this paper, coined the expressions 'A-series' and 'B-series'. We already know what these expressions refer to, so a further reiteration is not called for. Time involves change, says McTaggart. If nothing changed, there wouldn't be any time. This sounds all right until we wonder whether it is possible that everything now happening in the universe might spontaneously stop happening, and stay stopped happening for a certain period of time. Of course, whatever we would have used to time this period cannot be used for this purpose because it itself is numbered amongst all those things that have stopped happening. So if everything stopped for three days, the earth *would have* gone round on its axis three times, but that very process, which ordinarily would have given the measurement of three days, has stopped happening along with everything else. But even though there is a straightforward difficulty of seeing how the complete stoppage of all processes could be timed, it is not straightforwardly incoherent to say that everything happening might stop happening. It does look, on the face of it, that there would be a genuine difference between that case where all processes stop for a day, say, and that where they stop for a whole year. I do not really want to get into a debate about these ideas, because although they strike me as having a prima facie interest, they are not relevant to what I want to say about McTaggart.

Having said that time involves change, McTaggart goes on to say:

> If, then, a *B* series without an *A* series can constitute time, change must be possible without an *A* series. Let us suppose that the distinctions of past, present, and future do not apply to reality. In that case, can change apply to reality?

McTaggart thinks not. He is not willing to throw away change, so he feels obliged to reject the B-series. This is a mistake, and it follows as a consequence of his view about change, which is itself mistaken. Change, McTaggart believes, can occur only if there are tensed facts, only if events are past, present and future, because only then can events change their A-determinations.

McTaggart looks at the B-series and finds that change is absent. If any particular event has a position in the B-series, it always did and always will have that position. With respect to that, no change can occur. The same is true of this event's B-relations to other events. If it is the case that a certain B-relation holds between two events, that relation is permanent. There is no room for change here. The B-series is static. As McTaggart says, when we conceive of events arrayed along the B-series, we exclude the possibility of any event being at one time in the series, and at another time out of the series, and we exclude the possibility of any B-relations being sometimes thus, and being sometimes something different — if event A is three days earlier than event B, that is how things will stand for ever; though saying it like that might be misleading, because when we think of events arrayed in a B-series, we are not thinking of that from any particular temporal position. If it is true that A is three days earlier than B, that is not true at some times but false at others. We can only say something like the truth of 'A is three days earlier than B' is 'timelessly' or 'eternally' true. But I think I would prefer to say simply that that particular B-relation obtains. And 'obtains' is not a tensed expression when used in that way, even though it is the present-tense form of the verb. It is tenseless in the same way that 'adheres' is if we say that justice adheres in the hearts of good men. We do not mean that justice adheres only at certain dates — it's just a fact, so we are alleging, that good men are like that. Similarly, when we talk about B-relations, we are saying that the temporal relations between such and such events just are like what we say they are.

McTaggart now asks, 'What characteristics of an event can change?' He believes that there is just one class of such characteristics, namely their A-determinations. He offers the example of the event of the death of Queen Anne, pointing out that once that event was in the far future, at every subsequent moment it became less future until it became present, then past, where it will remain for ever, growing however more and more past. As we know, this sort of understanding of events might be making appeal to either the transient or the tensed view of time. McTaggart's concluding remark states:

> If there is no real *A* series, there is no real change. The *B* series, therefore, is not by itself sufficient to constitute time, since time involves change.

It seems fairly clear that McTaggart's understanding of these matters is hampered by his insistence on seeing change in terms of changes in events, and not as changes in things. If indeed the essence of time is change (which to my mind sounds altogether too vague to help very much with a philosophical investigation), as McTaggart apparently accepts, then that is so surely as a result of the changes that happen to things being in temporal relations, as well as some changes themselves having temporal parts, as when the change that a leaf in autumn undergoes starts with the leaf going slightly brown, and proceeds with the leaf turning golden, then yellow; plainly the process has earlier and later stages.

Events, to my mind, are not in that class of entity which can undergo changes. Of course, the event of my typing the first word of this chapter was once present, and now it is past — the A-statement 'His typing that word is present' has changed its truth-value from true to false, but even though that statement is now false, and it is correct to say that that event is now past, it doesn't follow that anything has actually happened to the event itself, or that the event itself has changed in any way. The incoherence of holding that the event is moving away down the river of time, or of holding that the event has shed the property of presentness and acquired the property of pastness, has been demonstrated.

Things change; when something changes, an event occurs. An event is constituted by something undergoing a change with respect to the states it is possible for that thing to be in. I would urge that the static view of time is quite capable of representing change, in which case, even if McTaggart is right in saying that change is the essence of time, we will find no good reason in connection with change for dismissing the static view of time.

Something changes when at one time it is in a certain state, and at another time it is in a different state. Thus, if a certain leaf is green at t_1 but red at t_2, the leaf has changed with respect to its colour. McTaggart is aware of this interpretation of change. He refers to Russell's *Principles of Mathematics*, section 442, where Russell says more formally what I have just written,

> Change is the difference in respect of truth and falsehood, between a proposition concerning an entity and the time *T*,

> and a proposition concerning the same entity and the time *T'*, provided that these propositions differ only by the fact that *T* occurs in the one where *T'* occurs in the other.

A change in the poker occurs when the proposition 'My poker is hot at T' is true, and the proposition 'My poker is hot at T'' is false.

I do not understand McTaggart's objection to this view of change. He correctly sees that Russell is looking for change 'not in the events in the time-series, but in the entity to which events happen, or of which they are states'. Now, what could be wrong with that? McTaggart is worried that if the poker is hot on a particular Monday, the event of the poker being hot does not change. But why does McTaggart want the *event* of the poker being hot to change? Events don't change, things do. If the poker was cold before it was hot, then at the time the poker is hot, it would be correct to say that the poker has changed with respect to its temperature. Its being true that the poker is cold at t_1, and its being true that the poker is hot at t_2, for McTaggart, seems to miss something out, and I am not able to understand his misgivings.

It seems to me altogether unproblematic to say that if we state in B-series terms that at t_1 the poker is cold, that at t_2 the poker is hot, then we have given the truth conditions for its being the case that the poker has changed with respect to its temperature. Even though the static view of time offers us just the B-series with all the events of history arrayed on it, permanently or 'timelessly' cross-referenced by a mass of B-relations, we have in that view all the room we need to see that things can change. If the B-statement 'Object A being green is earlier, by three days (say) than object A being red' is true, it is the case that A has changed from green to red.

McTaggart's wish to hold that real change is only change in events remains baffling. The static view of time seems quite capable of representing change, and I can see no need to smuggle in a dynamic analysis in order to make the account complete and coherent. The point to emphasise is that we can talk about change without the need to hold that events change their A-determinations, or that they are past, present or future, if by that we mean something more than that events are earlier or later than or simultaneous with other events, including the events of our thinking or talking about their temporal relations.

Note

1. 'Time'.

12
What Makes Tensed Statements True?

We have already seen, in Chapters 4 and 9, that the idea that A-statements or tensed statements which locate events on the A-series express tensed facts runs into difficulties; we know now that a tensed fact cannot consist in an event floating on the river of time, and it cannot consist in an event possessing temporal properties (of pastness, presentness or futurity). No other plausible candidate for what a tensed fact might consist in seems forthcoming. It is therefore a good thing that tensed facts are not needed to make tensed statements true. We demonstrate this all the time when we assess the claims that people make in tensed terms, and it is quite easy to see the truth of this matter.

To do this, we shall consider the simple tensed statement 'The bus left ten minutes ago.' The truth conditions for this statement can be expressed in tenseless terms: 'The bus left ten minutes ago' is true if and only if the following B-relation obtains — the leaving of the bus is earlier, by ten minutes, than the utterance of the statement 'The bus left ten minutes ago.' This is perfectly straight forward. The leaving of the bus occurs at a certain date, t_1 say, and the utterance of the statement 'The bus left ten minutes ago' occurs at another certain date, t_2 say. If t_1 is earlier than t_2 by ten minutes, the tensed statement about the bus leaving ten minutes before is true. The statement is true precisely because the bus did leave ten minutes before it was uttered, but these conditions, as we have just seen, can be expressed in tenseless terms. To express these conditions, we have merely to state the B-relation, 'The leaving of the bus is ten minutes earlier than the utterance of "The bus left ten minutes ago".'

Even for those who feel dubious either about the argument

offered in the first part of Chapter 9, or about its application to tensed statements under any possible interpretation as to what a tensed fact consists in, the understanding that tensed statements have tenseless truth conditions (rendered by expressing B-relations) ought to constitute sufficient grounds for accepting the myth of tensed facts. Were it not for the difficulties we have discovered, the one thing that a tensed fact would have been good for, would be to make a tensed statement true. But we don't need tensed facts to do this job, because tenseless truth conditions fulfil that function. Tensed facts are therefore logically redundant; we have good arguments (I believe) for saying that tensed statements do not express tensed facts, and that tensed facts do not make tensed statements true. To my mind, by way of example, this is perfectly analogous to the situation with respect to the nineteenth-century ether. The ether was supposed to be the medium in which electromagnetic radiation is transmitted, but a change in theoretical outlook made the ether redundant; it was conceptually redundant in that the transmission of radiation could be understood despite the ether. The only acceptable view is that the ether is a myth, it has no ontological status whatever. Similarly, since the notion of tensed facts is redundant even in the area where an application might be thought to have some initial plausibility, we can conclude that tensed facts are also a myth.

13
Can Tensed Statements be Translated by Tenseless Ones?

In this chapter I will show that A-statements cannot be translated into B-statements. It will help if we use ideas on this topic from D. H. Mellor's book,[1] and an instructive paper by Richard M. Gale.[2] In his paper Gale says that if tensed statements cannot be rendered by tenseless ones without loss of meaning, then the static view of time will be refuted. I believe that his arguments for concluding that tensed statements cannot be translated by tenseless ones are sound, and I accept that his criterion of meaning is sound; but I disagree, along with Mellor, that this entitles us to reject the static view of time.

When I say that an expression can be translated by another one, I would say that the two expressions concerned have the same meaning. And when I say that these expressions have the same meaning, I would say that they are logically equivalent, in that each expresses a common fact, or common facts, and that each can be used with the same success to express such facts, or to instruct or to order, or to achieve any of the functions that expressions are believed to have. Also, two expressions held to have the same meaning should not imply or entail different facts.[3]

Gale adopts as his criterion of meaning the use of an expression in certain contexts. He believes, correctly in my view, that tensed statements necessarily have a different use than tenseless statements, in which case tensed statements have different meanings from tenseless statements. Having discussed this, I will say in the next chapter why Gale is mistaken to further conclude that the static view of time is wrong; it is my belief that even though language cannot be 'detensed' (tenseless statements substituted for tensed statements) it does not follow that events are inherently past, present and future, that there are tensed facts which our tensed statements express.

It is easy to get confused about this issue because A-statements and B-statements can be closely connected. If the A-statement 'E is past' is true, then so is the B-statement 'E is earlier than the utterance of "E is past".' So, whenever 'E is past' could be truly asserted, so could the statement 'E is earlier than this utterance', where 'this utterance' refers to the statement within which it occurs. Thus, A-statements are token-reflexive; if true at all, they are true in virtue of their being uttered at particular times, analogous to statements which are token-reflexive in a spatial sense, such as 'Object A is here', which can be understood only if we know where in space the speaker is (and is true only if the speaker is in A's vicinity), and analogous to statements which are token-reflexive in a personal sense, such as 'I am cold', which can be understood only if we know *who* the speaker is (and is true only if the speaker is himself cold). Token-reflexive A-statements are not freely repeatable — they say something different, and have different truth conditions, every *time* they are used. That is, any token of such a statement in fact makes a different statement each time it is employed. Similarly with tokens of statements that are token-reflexive because they employ pronouns, such as 'I am cold', which makes a different assertion each time a different *person* utters it. And similarly with tokens which employ the token-reflexive terms 'here' and 'there' — 'Object A is here' says something different depending upon the *location* of the speaker. (My particular interest is in token-reflexive statements which lack free-repeatability because of the *temporal* terms they contain. I will seek to avoid confusing examples that have what can be called 'multi-reflexivity' such as 'I fell over that object there three days ago'.) Token-reflexive statements can be thought of as tokens of particular statement *types*. Thus, clearly, 'I am cold' when said by me is one statement, different in meaning and having different truth conditions from the statement 'I am cold' when said by someone else. And 'I am cold' when said by me at t_1 has a different meaning and has different truth conditions from the statement 'I am cold' when said by me at t_2. The first token means that my being cold is a state contemporaneous with its utterance, that is at t_1, and its being true is dependent upon the fulfilment of the condition that I am actually cold at t_1. The second token means that my being cold is a state contemporaneous with its utterance, that is at t_2, and it is true only if I am cold at t_2. The condition under which the token

'The bus left ten minutes ago' uttered at 2.10 p.m. is true, is that the bus referred to departed at 2.00 p.m., and the condition under which the different token 'The bus left ten minutes ago' uttered at 2.30 p.m. is true, is that the bus departed at 2.20 p.m.

Merely in noting that A-statements are token-reflexive, that A-statements are not freely repeatable, whereas B-statements are, appears to give us sufficient reason to conclude that there can be no B-statement which can translate a particular A-statement, because given the A-statement we can validly infer that it can be truthfully uttered only at certain times — this must be the case, regardless of the factual content of the statement, because A-statements are token-reflexive in virtue of their temporal terms alone, even if their reflexivity is compounded by other token-reflexive terms, as would be the case with the A-statement 'I sneezed yesterday.' But any B-statement offered as a candidate to translate the A-statement is freely repeatable, true whenever uttered. It cannot be the case that statement Y means the same as statement X, if in any contexts Y is uttered always saying something true, but in some of those contexts X when uttered says something false.

Another basic difference between A-statements and B-statements is that when a speaker utters an A-statement, he at one and the same time reveals his temporal relation with the event referred to in the statement. Thus, if S says truthfully that the bus left ten minutes ago, we know that the leaving of the bus and S's mention of it are temporally related, with the utterance of the A-statement occurring ten minutes later than the leaving of the bus. We know that for S, his perspective on what is happening places him ten minutes later in time than the leaving of the bus. B-statements tell us nothing at all about the temporal relations which obtain between someone who utters a B-statement, and the events he thereby talks about. If I were to say that the bus's leaving occurs ten minutes earlier than S's saying 'The bus left ten minutes ago', no inference can be made as to when, in relation to the bus leaving or S's mention of its leaving, I am saying that. This shows that no B-statement can be uttered which captures the whole meaning of an A-statement.

Someone who holds the possibility of translation thesis might say that the sole function of an A-statement is to express the B-relation that obtains between the utterance of the A-statement and the event the A-statement is about. This seems initially plausible, because, as I have pointed out, when the A-statement 'E is past' is true, so is the B-statement 'E is earlier than the utterance of "E is past'',' and that being so, it seems safe to say that what we understand by 'E is past' is that 'E is earlier than the utterance of ''E is past''.' What this shows is that given an A-statement we can derive a B-statement from it. This is just what we would expect, because the way in which we decide whether the A-statement 'E is past' is true, is to decide the truth of the B-statement 'E is earlier than the utterance ''E is past''.' That is, the condition under which 'E is past' is true, *is* E's occurring earlier than the utterance 'E is past.'

Mellor begins his analysis of the translation issue by suggesting that for many statements, giving a statement's truth conditions gives its meaning.[4] This explains the suspicion that a tenseless statement which gives the truth conditions of a tensed statement means the same as the tensed statement. Mellor offers a perfectly plausible example of this, pointing out that the statement 'X is half empty' is true if and only if X is half full; thus we can say that '. . . is half full' means the same as '. . . is half empty'. But even this straightforward example fails, I feel, because '. . . is half full' and '. . . is half empty' imply different facts. '. . . is half full' implies that the container in question, on this occasion, has not been more full than half full, whereas '. . . is half empty' implies that the container has been completely full, and that half the contents have now been removed. This usage might not hold strictly in every case, but in many contexts '. . . is half full' and '. . . is half empty' would imply different facts, if not in a strict logical sense, then in the sense of hinting at or pointing to a likelihood. And what an expression hints at is obviously part of its meaning — so two expressions which hint at different things cannot be said to have the same meaning. In any case, in looking at A-statements and B-statements and how they relate logically, we *en route* as it were establish the general claim that a statement which expresses the truth conditions of another statement need not necessarily mean the same thing.

The task in hand however is to find out whether the tenseless truth conditions for tensed statements mean the same as the tensed statements. If they do, tensed statements can be translated by tenseless ones, at least in the minimal sense that we could abandon using tensed expressions and get by just as well, meaning by what we say just what we would have meant if tensed expressions were still allowed, by using tenseless translations. That is enough to satisfy the claim that tenseless statements can translate tensed ones.

Mellor shows that the simple tensed statement 'It is now 1987' has no tenseless translation. We can summarise what he says by considering the following statements:

X: It is now 1987.
Y: X occurs in 1987.

Y gives the truth conditions for X. X is true, if and only if Y is true. Is Y a translation of X? If two statements mean the same, they must both be true under the same conditions, and there should not be conditions under which one can be true and the other false. X and Y are not both true under the same conditions — they in fact have different truth conditions. If Y is true, then any token of it uttered at any time would be true (Y is a B-statement, and this is true of any B-statement). But for a token of X to be true, it must be uttered in 1987 and at no other time. Since any token of the alleged translation of X is always true, but X itself is true only if uttered in 1987, X and Y cannot have the same meaning. We can see as well that X and Y entail different facts, and that being the case again establishes that they cannot mean the same thing. If we hear a true token of X we can infer that the year is 1987; but if we hear a true token of Y, we can make no inference either way as to whether it is, or is not, 1987.

It will help if we look further into this by considering what Gale says in his discussion of a practical situation where someone is using an A-statement to convey information. We can then decide whether there is any way to convey the same information by speaking tenselessly. We will see that there is not. Gale uses the following example:

> Joe is a scout for a machinegun company. He is strategically stationed so that he can survey the battlefield, and when the enemy approaches within 100 yards of their position he must inform the company so that they can open fire.

What we want to know is whether Joe can inform the company using a tenseless expression as well as through the use of a tensed expression.

The usual way for Joe to inform the company would be to utter the tensed statement 'The enemy is now within 100 yards.' This statement has tenseless truth conditions. It is true if and only if the enemy's being within 100 yards is (tenselessly) simultaneous with

Joe's utterance of 'The enemy is now within 100 yards.' The condition here is expressed as a B-statement, and if true, is 'timelessly' true. But the condition, in itself, does not capture the meaning of the original A-statement — in stating the condition, one says something that has a different meaning from the A-statement. This is clearly seen if we apply Gale's criterion of meaning (which I accept as a satisfactory criterion) which says that expressions are equivalent in meaning, under a specific context, if they have the same use. We are to imagine that when Joe sees the enemy approach, he says to himself 'The enemy is now within 100 yards', and then he 'translates' this into a tenseless B-statement which he says aloud for the company to hear, 'The enemy's approaching within 100 yards is (tenselessly) simultaneous with the utterance of "The enemy is now within 100 yards".' Gale points out that in saying this to the company, Joe is just mentioning the tensed statement, but is not using it. The outcome is that since the company did not hear the original tensed statement, they are not informed about the present approach of the enemy. What Joe says out loud for them to hear is indeed true, but unfortunately for the company, what Joe says would be true no matter when uttered. It so happened that Joe uttered the tenseless expression just after he said the tensed expression to himself, but the company have no way of knowing that. For all they know, the tensed expression which they hear Joe mention was uttered several days ago.

A further observation to make about this would be to criticise Joe's whole technique. What we were after was a translation of the original tensed statement. We wanted just that, not a translation plus the original tensed statement. If one expression is genuinely a translation of another, we must be able to use the translation on its own, and do the very same job with it that we would have done with the expression it is a translation of. What happens if Joe utters *just* a tenseless statement? Well, in a way, this cannot be done at all. Let us imagine that Joe, when he sees the enemy approach, simply utters aloud to the company the B-statement 'The enemy's approaching within 100 yards is (tenselessly) simultaneous with the utterance of "The enemy is now within 100 yards".' This will not do, because Joe is now mentioning a tensed statement which in fact does not occur — and if it does not occur, there can be no B-relation between it and another event to express. That difficulty aside, the company would still not know whether to open fire or not.

Joe might try mentioning an utterance which does actually

occur. When the enemy approaches, he might say 'The enemy's approaching within 100 yards is (tenselessly) simultaneous with this utterance.' Here, the utterance mentioned is the utterance which does the mentioning — in other words it is token-reflexive, and it is thereby not freely repeatable, and it is not a tenseless statement true no matter when uttered. This expression means something different each time a token of it is used. It *is* a translation of the tensed statement 'The enemy is now approaching within 100 yards' — and the company would know that they ought to open fire — but it is not a tenseless translation.

There is another way in which language can be 'detensed'. This involves expressing the 'timelessly' true B-relation which obtains between the event referred to by a tensed statement and that which serves as the origin of the calendar we are using. In other words, we can give the date of the event in question. To do that is to make a 'timelessly' true tenseless statement. If we say 'Such and such occurs on 14th May 1987', we are saying something equivalent in meaning to 'Such and such occurs 1,987 years four months and fourteen days later than the birth of Christ.' A listener will only know that the event referred to is happening in the present if he knows the date. 'Such and such occurs on 14th May 1987' is equivalent in meaning to 'Such and such occurs now' only because the 14th May is the current date. 'Such and such occurs now' is not freely repeatable. Said tomorrow, it would state something false. To say truthfully that a certain event occurs at a certain date is to say something that is freely repeatable — the statement used to do that is tenseless. Saying that something occurs now is obviously to use tensed language.

We can imagine that Joe persists in trying to get a tenseless translation for 'The enemy is now approaching within 100 yards' by stating the date at which the enemy approaches. Gale points out that Joe would need a watch which records the date as well as the time of day. We will grant him that. He sees the enemy coming, looks at his watch, and says 'The enemy approaches within 100 yards on 14th May 1987, at 4.27 p.m., BST.' This is an improvement, because this is a genuine B-statement, and if the company knows the date and the time of day, it will know to open fire when Joe makes the statement. But this supposed translation fails on two counts. Firstly, 'The enemy approaches within 100 yards on 14th May 1987, at 4.27 p.m., BST' does not entail that now is the right time to open fire; Joe can utter a token of that B-statement an hour (say) after the approach of the enemy, and say something true. It would not however be right to infer that the enemy is approaching. But if Joe truthfully uttered a token of the tensed statement it is supposed to translate, namely 'The enemy is now approaching within 100 yards', it would be right to infer that the enemy is approaching. Therefore, a token of the tensed statement cannot mean the same as a token of the tenseless statement. And secondly, the point Gale wishes to emphasise, the two statements do not have the same *use*. Joe uses the tenseless 'translation' successfully only because he and the company possess accurate watches by means of which statements which relate events to dates can yield information as to whether the events are happening now, or at some other time. Gale argues in terms of Wittgensteinian language-games, pointing out quite rightly I think, that when Joe uses the tensed expression one particular language-game is employed, and when Joe uses the tenseless expression a different language-game is employed. From one language-game to the other, the information that the company needs to know, is conveyed by quite different means. As Gale concisely sums up:

> In order for the tenseless way of speaking, in which dates are ascribed to events, to work in such contexts, it would be necessary for the people to know the exact date and time, but to know this requires additional effort on their part, so that the two modes of talking are not both equally legitimate and workable.

Gale stresses his claim that non-equivalence of use must mean non-equivalence of meaning, by concluding with a humorous example. We have to imagine the one case where at an appropriate point in the proceedings a woman says to her lover 'Kiss me *now*!', and another case where the woman says 'Kiss me at such and such date!' In the second case the lovers would have to put the light on to consult their watches for such an instruction to result in the requested action. Gale is concerned that the consequent lapse of time might prove disastrous. As far as engaging in the sort of practices that people do in fact engage in, Gale's point seems beyond objection. We can add to this claim that even if we could imagine beings engaging in practices the proper working of which never required the use of tensed statements, in which case they would be able to stick exclusively to the use of tenseless statements, such an example would be redundant to the issue of possible translation, because in this case the question of translation doesn't even arise.

The point is that we need go so far as making our philosophical conclusions consistent with actual human behaviour and no further.

The overall findings presented in this section, about truth conditions, entailment relations, and the practical use of tensed expressions, point to the satisfactory conclusion that tensed statements cannot have tenseless translations.

Some writers, including Gale, think that this defeats the static view of time. I will show in the next chapter why I do not think that that is so.

Notes

1. D. H. Mellor, *Real Time*.
2. Richard M. Gale, 'Tensed Statements'.
3. In this chapter I use these terms in this way: if A entails B, it is correct to infer B from A.
4. See D. H. Mellor, *Real Time*, pp. 73ff.

14
Tenselessness

Given the occurrence of a true token of an A-statement, a true B-statement can be derived: for example, 'The bus left ten minutes ago' (A-statement) yields 'The event of the bus leaving is ten minutes earlier than the event of the token "The bus left ten minutes ago" ' (B-statement). But given a B-statement, we cannot derive an A-statement which says whether the events mentioned occur in the present, past or future (or how far in the past or future). Given 'Event A is earlier than event B', we cannot validly infer any A-statement. Thus to say one or some of 'A is happening now', 'B will happen', 'A is in the past', 'B is in the past', or whatever, may be to utter true A-statement tokens. But knowing just that the given B-statement about A and B is true gives too little information for us to decide which possible A-statement tokens could now be uttered truthfully.

This summarises one important finding that arose when seeing in the last chapter whether tensed statements have tenseless translations. Gale sees this as a sufficient reason to reject the static view of time:

> Since tenseless statements can be derived from tensed ones, but not *vice-versa*, this shows that the *A*-determinations are fundamental and that *B*-relations are reducible to them, but not *vice-versa*. This establishes the validity of Broad's analysis of the concept of time [a tensed analysis].[1]

I think Gale is wrong. We already have good arguments for dismissing tensed facts as a myth; contrary to what Gale urges, events are not inherently past, present or future. My aim is to show in this section that talking in tensed terms, indeed *having to* if we are to continue our practices successfully, is perfectly consistent with its being the case that time is tenseless. This will be done by showing that, with respect to at least tense, time and space are analogous. Space is tenseless and so is time. How extensive the overall similarity is between space and time is not a question that needs to be answered for our present purpose. Gale's position about time rests on the general claim that if we are talking in tensed terms, then what we are talking about is tensed. This general claim can be refuted by showing that it does not follow from the fact that we talk in *spatially* tensed terms that space is tensed. If it is not true that tensed talk invariably points to tensed facts, it does not directly follow from the fact that we use temporally tensed language that time is tensed. I am not sure that there is an argument which shows that time *must* be tenseless *because* we use temporally tensed language. But that doesn't matter, because I do not require such an argument. Since good arguments can be found which refute the notion of temporally tensed facts, I will do enough if I can show that tenseless time is consistent with tensed temporal language.

The A- and B-series of time have direct analogues with respect to space. Just as the A-series relates all events and dates to 'now', the spatial A-series (which I shall henceforth indicate by subscript, A_s) relates all objects and places to 'here'. Objects and locations thus have A_s-determinations: for instance, since I am here, London is 15 miles to the south-east. Thus, the spatial directions of north, south, east and west are analogous to the temporal directions of past and future. Just as any event or date must occur in either the present, or at some temporal distance past or future *from now*, any object or place must be either here, or some distance at some cardinal point *from here*. Temporal distance is expressed in terms of days, seconds, years, centuries, or what have you. Spatial distance is expressed in terms of miles, inches, yards, or what have you. For clarity and ease of comparison, since I am referring to spatial A-determinations as A_s-determinations, I shall (in this chapter) denote temporal A-determinations by using another subscript, and call them A_t-determinations. As we know, A_t-statements have tenseless truth conditions. So do A_s-statements. The A_s-statement, 'London is 15 miles south-east of here' is true if and only if London is 15 miles south-east of the place where it is uttered. Any token of that A_s-statement is not freely repeatable: if spoken saying something true, it must be spoken here, where I am, because this is the place from which London is 15 miles to the south-east. But any token which expresses the truth conditions of our sample A_s-statement, viz., 'London is 15 miles south-east of the place where "London is 15 miles south-east of here" is uttered', in the event that someone does make the A_s-statement here, is true at any place it is uttered. We can thus see that spatially tensed statements and their spatially tenseless truth conditions operate analogously to the way that temporally tensed statements and their temporally tenseless truth conditions operate.

We can see as well that the analogous relationship applies similarly to B_t-relations and B_s-relations (where the subscripts 't' and 's' again mark the senses of 'temporal' and 'spatial'). Just as events and dates in time can be conceived as arrayed along the B_t-series, objects and places can be conceived as arrayed across a one-, two-, or three-dimensional B_s-series. That space has more dimensions than time complicates the way in which objects and places can be spatially related. This does not affect the claim that space and time are analogous in the sense that each is tenseless, which we are now investigating. Directions in time number two — earlier than, and later than. There are many more directions in space, since the basic directions of north from, south from, west from, east from, up from, down from, can be combined to indicate an infinite number of directions that can be taken from a particular point in three-dimensional space. (Here is a blatant example of one way in which space and time differ — if object A and object B are spatially related, object B can occupy any one of an infinite number of positions which lie in an infinite number of different spatial directions from A. Whereas if event A and event B are temporally related, B can only be either later than A, or earlier than A: there are just two temporal directions, but an infinite number of spatial ones.) That the arrangement of objects in space is more complicated than the arrangement of events in time does not undermine the claim that both space and time are tenseless.

Thus a map of part of the earth's surface represents graphically the B_s-relations of such things as cities and mountains and rivers. The B_t-statement 'Event A is ten minutes earlier than event B' is logically analogous to the B_s-statement 'Object A is ten miles to the north of object B.' If true, tokens of those statements are freely repeatable: the B_t-statement can be uttered at any *time*, and the B_s-statement can be uttered any*where* (or more comprehensively, each can be uttered by anyone, at any time, in any location).

The system of longitude and latitude works like the temporal calendar, although to locate places on the earth's two-dimensional surface, reference to two origins is necessary, those being the equator and the Greenwich meridian. Accordingly, any object or place has such and such distance from the equator and such and such distance from the Greenwich meridian. Similarly, any event or date is such and such temporally distant from the birth of Christ.

With respect to time, an event or a date can be located on the B_t-series by relating it to some other event or date, or by giving its own date. If referring to the invasion by the Normans, we can say that the Norman Conquest occurred nine centuries earlier than the Second World War, or we can give its date by saying that it occurred in AD 1066. Similarly, with respect to space, an object or place can be located on the B_s-series (which is just a technical way to say that objects and places can be located in 'space') by relating it to some other object or place, or by giving its spatial co-ordinates. If referring to Watford, we can say that Watford is located 15 miles north-west of London, or we can give its co-ordinates by saying that Watford is at 51° 39′ N, 0° 23′ W.

When we utter B_t-statements, or give the dates of events, we say nothing about our temporal relation to those dates or events. And similarly when we utter B_s-statements or give the co-ordinates of objects or places, we say nothing about our spatial relation to those objects or places.

Having drawn out the analogy between space and time, we can now proceed with the discussion. Since it is true, (as we shall see) that the fact we talk in spatially tensed terms does not entail that space is tensed (meaning by that that there are no spatially tensed facts) it follows that the claim to the effect that, if we are talking in tensed terms, then what we are talking about must be tensed, is false. This is a refutation by counter-example, since we know it to be true that we talk in spatially tensed terms, yet space is not tensed (that is, there are no spatially tensed facts) — as I shall now indicate.

No one believes in spatially tensed facts. It is not an inherent fact about Watford that Watford is *here*. It is not an objective fact about reality that London is 15 miles south-east of here. No objects are really *there*, or really 'nearer than' other objects — although they are spatially related. The tree yonder does not have the objective property of being yonder, which it somehow loses to acquire the property of being here as I walk towards it. That Watford is north-west of London is not made the case by Watford's mysteriously possessing the property of 'north-westness'. To hold that there are spatially tensed facts, we would have to say that Watford also possesses the property of 'southerliness', for this is the tensed fact which on a tensed view of space would make the spatially tensed statement 'Watford is south of here' true when uttered in St Albans. But since we know that Watford is spatially related to all the places there are in Britain (and beyond) a tensed view of space would require that Watford possess that property attaching to each and every cardinal point in order to *be* related to all these places. This is more than absurd, because no place could equally possess the properties of 'being south' and 'being north' since these 'properties' are incompatible. If it is ever correct to say that a certain place is south, it is necessarily false to say that it is also north.

Clearly there are not spatially tensed facts, but there are tokens of spatially tensed statements, if we care to make them, which are made true, if they are true, by spatially tenseless truth conditions. It plainly does not follow from the fact that we do not use spatially tensed statements that space is tensed. Similarly, bearing in mind the logical analogy that we find to hold between space and time, the mere fact that we do make use of temporally tensed statements does not entail that there are temporally tensed facts.

The tenselessness of both space and time is further illuminated by noting how we use the key tensed expressions 'here' and 'now'. This will involve reiterating some of the points already mentioned, making for further clarity. One fact about the sort of creatures that human beings are, is that at all points in our lives, whatever is going on, whatever we are doing or observing, is taking place here and now. Each person finds himself at the 'centre of things'. As far as each individual is concerned, things happen in relation to his 'here' and his 'now'. In this sense, each person finds himself at the focus of events, the centre of experience.

Thus, we need to be able to say when things are happening in relation to our experience — whether they are simultaneous, or earlier than or later than our experience, and often how much earlier or later than. We do this by using tensed, token-reflexive, language. For example, the A_t-statement presented on a card behind a shop window 'Back in ten minutes' conveys information about what is happening only if we know the time at which the notice was pinned up, and if we know the present time. That expression is token-reflexive because it makes an implicit reference to itself — it means 'I will return ten minutes later than the time at which this token was pinned up.' Similarly, we understand the A_t-statement 'Event A is now happening' because we know that it means that event A is simultaneous with the utterance of the token itself, and with our hearing it.[2] Thus, if we hear the token, we know that event A could in principle be experienced (it may not of course be happening in our spatial vicinity). That our language operates in this way is not sufficient grounds to conclude that event A really *is* now in the way the tensed theorist wants us to understand things.

Event A is no more *really* now than object B (this typewriter, say) is *really* here. The analogy between space and time is further emphasised when we see that we need to say *where* things are in relation to our experience — since our experience is where our bodies are (and that appears to be a contingent fact about human beings that need not have been the case) saying where things are in relation to our experience amounts to saying where things are in relation to our bodies. 'Here' is wherever my body is (depending on the context; 'here' might refer to a small area of space, 'Here, in the spot where I am standing', or a slightly larger area, 'Here, in this house', or larger still, 'Here, in Britain').

In saying that object B is here, we have no inclination to believe that object B really *is* here, has 'hereness', or is really spatially tensed. Its 'being here' consists only in the tenseless fact that its spatial co-ordinates are suitably similar to those of my body (bearing in mind the context and my aims in judging or saying that object B is here). Those are the spatially tenseless truth conditions for the tensed statement 'Object B is here.' Similarly, for other tensed expressions. If object B is nearby, that is not in virtue of some quality it has. The tensed statement 'Object B is nearby' is 'dynamic' in that it changes its truth-value depending upon where it is uttered. But there is no corresponding property which B gains and loses as a consequence of my moving about, nearer to and further away from, B. The spatially tensed statement 'B is nearby' does not change its truth-value because of anything happening to B (such as its changing its A_s-determinations), nor because of anything happening to me. It changes its truth-value only in so far as the spatially tenseless truth condition 'He is within such and such distance of B and no nearer' obtains. 'Nearby' operates analogously to 'recent' and 'expected'. Event A is recent only to the extent that certain temporally tenseless truth conditions obtain.

'There is something unsatisfactory here,' it might be remarked.

'The idea that events do not "really" happen in the present (or at other times), and the idea that an object cannot really be "here" seem puzzling. In a perfectly ordinary straightforward sense people often talk about events happening now, yesterday, next week, or whenever, and people often talk about objects (and people) being here or there. Everyone understands what is being said when such things are asserted, and again, in a perfectly ordinary straightforward sense, everyone knows whether statements about such matters are true or false when they are encountered in daily life. To say that the chair I am sitting on (for instance) is not "really here" seems straightforwardly false. Here is where the chair is, and sitting on it is where I am.'

There is no problem here. When the speaker says 'The chair is here' his statement is made true by the spatially tenseless condition of his body being appropriately close to the chair: if he is sitting in it, he and the chair would be in contact. His tensed statement is true because a tenseless state of affairs obtains. This is the point to emphasise — tensed statements (spatial and temporal ones) are of course true or false, but the conditions under which they are either true or false, are tenseless. People in everyday situations using language in an ordinary way do not need to be aware of this — tensed statements in everyday discourse are treated as if they express tensed facts, and as if tensed facts are what satisfy their truth conditions. But in metaphysical discourse about time,[3] to say that an A-statement is true implies that the fact which it seems to express (a tensed fact) obtains. But there are no tensed facts. In saying that an A-statement is true, without saying everything else we need to say about tenseless truth conditions and the myth of tensed facts, is (in the metaphysical sphere of discourse) misleading.

Having argued that time is tenseless, that there are no temporally tensed facts, it is comforting to see that this fact is consistent with the fact that we talk about events and dates in temporally tensed terms. Tensed language points to tensedness no more than it points to tenselessness, since space is tenseless, but we talk in spatially tensed terms about objects and places.

Notes

1. Richard M. Gale, 'Tensed Statements', p. 59.
2. Allowing, that is, for the speed of sound, and the time it takes for

brains to process tensed statements, and voices to speak them.

3. And similarly with respect to space.

15
Dynamism Denied Outright

It might be said that if events cannot have temporal properties and therefore cannot *change* their temporal properties, if time cannot be dynamic in that sense, then time is dynamic in at least the minimal sense that A-statements change their truth-values. Thus the A-statement 'The Second World War is future', was at one time true, but is now false — it has therefore changed its truth-value.

This view is wrong. Time is not dynamic in even this minimal sense. This follows straightforwardly if we accept that it is not a statement in itself which is true or false, but an assertion, made by someone at a particular time by uttering a token. If someone utters the token at t_1 'E is in the future', that token will assert something true if and only if E occurs later than the utterance of the token. We can imagine the token being uttered again at t_2, a time later than the occurrence of E, in which case whoever used the token on that occasion would have asserted something false. Here we have two assertions, each made by one of two tokens, uttered at two distinct times: one is true, and the other is false. Both assertions happen to have been made using exactly similar tokens, and when we note that, we see that there is no single A-statement which has *changed* its truth-value.

The form of words, the type, 'The Second World War is future' (which we cannot actually examine — that was just a token of it), is neither true nor false. The question of truth only arises when someone utters a token of the type on a particular occasion for the purpose of making an assertion. That assertion is either true or false. The same statement can be used on different occasions, but the same thing would not thereby be asserted. A token of 'The Second World War is future' might be uttered in 1935, and another token of it might be uttered in 1945. Clearly it is not the case that the same assertion is being made in both cases. In the first case, the speaker is asserting that the Second World War will occur later than the time of his utterance, which occurred in 1935, whilst in the second case, the speaker is asserting that the Second World War will occur later than 1945, the time of his assertion. Here, tokens of the same statement type are being used on two separate occasions to make two distinct assertions. As it happens, the first is true and the second is false.

This analysis[1] of statement types, tokens, and use of tokens to make assertions on particular occasions, appears to be perfectly consistent and coherent. To my mind it illustrates a sensible application of Occam's Razor, since in accepting it we are maintaining that the world does not contain a species of entity members of which can change their truth-value. This seems much more satisfactory than holding to the converse, whereby we would suppose that those things of which logic treats which 'have' truth-values are not true or false 'timelessly' or 'eternally'. An assertion uttered on a particular occasion occurs at only this date, and so long as it is meaningful, so long as it makes a claim that could logically be the case about something which can be a legitimate object of reference, then that assertion is either true or false, and whichever it is, it is so 'timelessly'.

If we accept this analysis, then A-statements (in particular) cannot *change* their truth-values, because statements in themselves are not true or false, although a token of any may be used to say at one time something true, and at another time something false.

Note

1. See especially P. F. Strawson, 'On Referring'.

16
A Note on Usage

That I shall continue to use tensed terms, and make use of examples in which subjects think of events as lying in the past or future, or in the present, should not be taken as evidence that really I am thinking of time transiently or as being tensed. As I have tried to urge in Chapters 12, 13 and 14, even though time is tenseless we can (indeed, sometimes have to) refer to events as being past, present, or future, even though they are not really past, present, or future. That we can say meaningfully, and truthfully, that E is past, does not entail our statement's being made true by the tensed fact of E's pastness. If it is ever correct to utter a tensed token, that is only because certain tenseless conditions obtain.

For instance, I believe that my typing the next page is future, but unlike the dynamic theorist, I hold that what this belief really comes to is the belief that a certain B-relation obtains, namely, that the occurrence of my thought about typing the next page occurs earlier than (to whatever degree) the event of my typing the next page. The thought about typing the next page lies at no temporal distance from my current experience, because it *is* my current experience. The actual typing lies in the direction 'later than'. It doesn't lie later than the tensed 'now', merely later than my thought about it. This analysis of my tensed belief makes no claim upon tensed facts.

Similarly, in continuing to talk about the past and future, I am not covertly smuggling in tensed facts behind my own back.

Part Two

17

Introductory — What our Experience is Like

The way we feel about our experiences, the view we have of what it means to be a creature who lives a life, tends to make us think certain things about the nature of time, of the past and the future, of our deliberations and strivings. There appear to be a number of facts which describe what our experience of living a life is like, which point to several asymmetries between the past and the future; there appear to be features about the past which are not true of the future and vice versa. It strikes me as a very obvious point, a point I intend to work out from, that we feel very differently about the future than we do about the past. My main aim in Part II will be to examine some of these differences in order to see which of our intuitions about time are correct, and which wrong.

The facts about our experiences, which exemplify our different attitudes towards the past and future, in which I am interested, are these. However we decide to analyse the sentence, there is a clear sense in which our lives are oriented towards the future. We act in the present to determine what will happen, but never act towards the past. The past is fixed and done with; there is nothing that can be done to make any of it into something different. It can be studied, remembered, talked about, but it cannot be tampered with. This is in complete contrast to the way in which most people regard the future. The very essence of the future is its vulnerability to our actions. If it is 'there' at all, it seems to be there in a hazy sort of way. If we wish to say that the past exists, most of us would feel inclined to say that the future, if it has any existence, has a very different sort. Whatever we say about that, it seems unquestionably true that our present actions determine what the future will be like, but they do not determine what the past was like. The various interpretations of fatalism contrast with all these intuitions, including even the final one, that our present actions determine the future, which the fatalist would say was not unquestionable at all, but wrong; our present actions determine what the future will be like to the same extent that they determine what the past was like, that is, they determine nothing at all — what will happen will happen no matter how much we try to influence events; the future, like the past, is fixed. Another brand of fatalism would argue that what we do does influence later events, but it's just that what we do has all along been fated to occur, so similarly we cannot change anything, or aim to bring about what we want, because our aiming is no more under our influence than events themselves.

Another fact which underlies the asymmetry between past and future is that the thought we have that some desires are *going to be* frustrated is much more disturbing than our knowledge that similar desires *have been* frustrated. The anxiety that we extend to a painful experience while the event of that experience is in prospect is altogether different from the anxiety we may, or may not, extend to the same painful experience once it has been suffered. It does not seem possible to be concerned about a past unpleasant visit to the dentist (if there is concern here at all) in the way in which it is usual to be concerned about a similar visit we anticipate making tomorrow, or some time in the near future. The fact that we cannot now do anything to avoid a past painful experience will obviously be intimately involved in explaining how our anxieties differ depending upon whether they are forward-directed or backwards-directed. The fact that we are concerned in this way when we anticipate an unpleasant future event, on Schlesinger's view, only makes sense if we accept that events really do move on the model of what I have called the transient view of time. I think Schlesinger is wrong in this; I have already shown that both the transient and the tensed conceptions of time are logically incoherent, but I think that our being concerned in the way we are makes sense within a static conception of time in conjunction with an understanding of how our acting relates to our feelings about what is happening to us.

The fact that we feel relief when an unpleasant experience finishes, such that it is appropriate to exclaim something like 'Thank goodness that's over', might seem to support the tensed view of time. On the face of it, when someone thanks goodness that his headache, say, has just ceased, he is thanking goodness for the tensed fact that the headache is past. I intend to show that there is an account of this sort of relief which does not require tensed facts and which does explain why we feel relief at the times we do.

Closely related to our being concerned about anticipated unpleasant experiences is our concern about our own eventual deaths. The time later than our deaths is a time during which we shall have no existence, just as the time before we were born was a time when we had no existence. This leaves us with a need to explain, if we can, why the thought of non-existence after our deaths pre-occupies us as it does, whereas the thought of our non-existence before our births does not. It looks as if one particular future event can threaten someone much more than any other conceivable event — that is the event of their own death. There are no past events that threaten us at all, or if we do decide to accept that recollecting some past events can trouble us or weigh us down, or something of the sort, their recollection quite clearly cannot concern us half as well as the event of our eventual deaths. Our deaths, as well as some other future events, can deprive us of those things we value; past events have no such power. Since we are either alive, in which case we can have experiences, or we are dead, in which case we can have no experiences, it appears that we cannot experience our own deaths. Of course we could experience pain, if for instance, we were dying from a painful illness, or had suffered an accident, but what we cannot experience is the passing away itself. It is not at all clear that death is something we are capable of suffering. If death deprives us of anything, then where is the person who has been deprived? Nagel's view on this[1] is that suffering need not in all instances be dateable, which he tries to show by considering some examples where we feel certain there *is* a victim who has suffered, or has been deprived, somehow or other, but that, as in the case of death itself, it seems impossible to find a person who suffers the deprivation at a particular date.

It is important to decide whether or not it is rational to fear death — not whether it is rational to fear the harms that one's death can undoubtedly bring (such as unhappiness for friends and family, and other harms), but whether one can rationally fear non-being, not existing. One has to be alive to sustain losses (such as bodily injury, or robbery), but can death itself be a loss? Being alive is a prerequisite of gaining benefits and sustaining losses, so it's hard to understand that the opposite of being alive (being dead) is in itself a loss.

Note

1. In 'Death'.

18
Concern about What is to be Experienced

I shall begin my discussion of our differing attitudes towards past and future by examining one particular feature of our experience which contributes to these differing attitudes. It is something which we feel almost every day of our lives. This is the fact that the prospect of an unpleasant experience is very commonly distressing to us; whereas the recollection of an unpleasant past experience, even one that is quite recent, does not arouse the same level of distress, and I think we can ask further whether, if indeed it arouses any distress at all, it is the same sort of distress as that created in us by the prospect of an unpleasant future experience. It would be a very great blow if a philosopher were to establish that our feeling about this is unwarranted — and if we felt convinced by such findings and lived our lives differently, I think we would be very different people from how we in fact are.

It will be useful to utilise the example which Schlesinger selects in his discussion of this matter, since it will be easier to address ourselves directly to Schlesinger's thoughts via the example.[1] We can imagine Solzhenitsyn's character Ivan Denisovitch being told that he is to serve fifteen years in a prison camp. Just as we all would be, Ivan is profoundly distressed at the thought of having to endure the next fifteen years of his life in the harsh and unpleasant environment of the prison camp. The question is to discover how this fact about what human beings are like can be fitted into our overall theory of time. It is Schlesinger's belief that Ivan's distress makes sense only if the transient view of time is true. Since the transient view of time is wrong, either Ivan's distress makes sense on another conception of time, or it doesn't make sense.

Before we continue, let us be clear about what Ivan's situation amounts to, making use of this diagram:

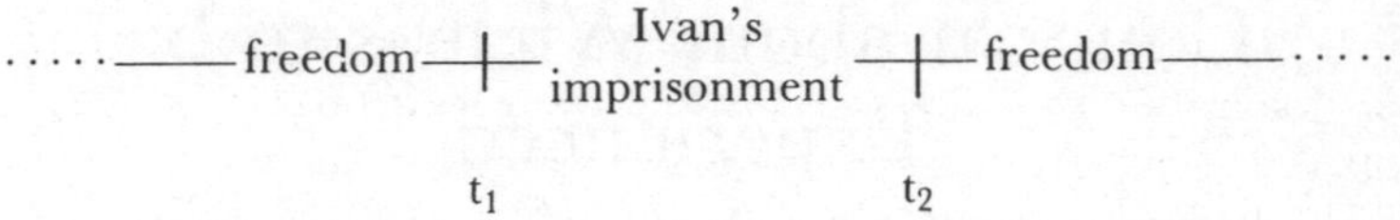

$$\cdots\cdots\text{——freedom}\!\!-\!\!\!+\!\!\!-\ \underset{\text{imprisonment}}{\text{Ivan's}}\ -\!\!\!+\!\!\!-\text{freedom——}\cdots\cdots$$

$$t_1 \qquad\qquad\qquad\qquad t_2$$

If we take this diagram to represent Ivan's situation *statically* we see that at times earlier than t_1 Ivan is a free man, but at times later than t_1 but earlier than t_2 he is not; there is a period of freedom which precedes a period of imprisonment. At times later than t_2, Ivan is free, but at times earlier than t_2 but later than t_1 he is not; we see a period of freedom occurring later than a period of imprisonment.

Ivan is told of his prison sentence at t_1 just as it is commencing. Is it true that on the static view of time Ivan's misery is a complete mystery? What reason has he to feel unhappy? After all, at t_1 he is very close to a period of freedom, which is perhaps just hours or days away. Schlesinger suggests that the reason why the period of freedom earlier than t_1 cannot relieve Ivan's distress is because Ivan is not *moving* in that direction. He is, unfortunately, *progressing away* from that period of freedom and heading *into* a period of unpleasant imprisonment, which he will have to endure until he *enters* the period of freedom which lies after t_2. 'Naturally, what determines his mood is the distance between him and the exit through which he will eventually pass, *leaving* behind the sorrowful times he is now enduring.'[2] Schlesinger says that if we are to understand Ivan's unhappiness, we have no choice but to conceive the diagram *transiently*, such that Ivan travels along the line from left to right.

Arguments have been offered in Part I to reject Schlesinger's view of Ivan; there is no need to rehearse them here. Time is *not* transient in the way that Schlesinger believes it is: there is no *entering* and *leaving* — Ivan's period of imprisonment is not approaching him, and his happy days of freedom are not receding away from him. If we agree with this, and I have already given my reasons in Part I why I think we should, do we after all have no means to explain Ivan's despair? Schlesinger thinks that his position is especially strengthened by the fact that a static theorist, in a position similar to Ivan's, would feel similar distress, revealing that such theorists cannot really believe the conclusions of their own analyses. Is then the fact that Ivan is miserable at t_1 explainable only by supposing that is is *moving* towards the period of imprisonment? Is it really true that Ivan would have no reason for despair were he in fact not moving in this way?

Schlesinger's argument is of course quite general. These feelings of despair which Ivan has illustrate the general fact that people often feel differently towards their past experiences than they do towards experiences they anticipate having, which I have already outlined in the previous chapter. So our discussion of Ivan's particular situation and feelings is really directed at the wider question as to whether the actual facts about people's attitudes (of the sort we are here interested in) can be explained only if we think of time as transient. Schlesinger's argument maintains that Ivan's attitudes about his future experiences, whatever in fact they may be, is *explained* by the fact (if it is one) that Ivan is *moving towards those future experiences*. So on this view, the only plausible reason that someone in Ivan's position can offer to account for his distress at t_1 is that he is heading towards a period of unpleasant experiences, or that the unpleasant experiences are bearing down upon him, and will overwhelm him, just as in a spatial context an approaching army can overrun the defenders.

Let us be quite clear about what Schlesinger's position is. He is not saying that Ivan's attitudes are caused by the unpleasant experiences he will be having later on; the unpleasant experiences which occur within the period t_1 to t_2 are not taken as *causing* his distress at t_1, and neither are the pleasant experiences which he will have in his state of freedom after t_2 taken as causing his supposed happiness at t_2. Ivan's attitudes are the result simply of his temporal movement towards his future experiences. It isn't perfectly clear, but I don't think Schlesinger can believe that this supposed temporal movement *causes* Ivan's attitudes, because there is an obvious difficulty in understanding how this movement, taken simply as movement, can cause different attitudes at different times. To account for Ivan's depression at t_1 and his happiness at t_2, the theory needs to make reference to those experiences he is moving towards, in which case it cannot be claimed that the movement, of itself, causes Ivan's different attitudes. So quite what the term 'result' can be taken to signify in the claim 'Ivan's attitudes are the *result* simply of his temporal movement' is obscure. The move to accepting that the future experiences which Ivan will later have, themselves partly (or even wholly) cause his earlier attitudes, is to be resisted for the obvious fact that his distress would be of exactly similar character in the one situation where he *is* truly convicted and as a matter of fact *does* live through an unpleasant period of imprisonment, and also in the other situation where the whole business proves to be a joke, and he in fact *does not* live through such an unpleasant period, but in fact lives a period of largely happy experiences — in this second situation we see that there are in fact *no* unpleasant experiences to cause Ivan's earlier distress, which the amended theory seems to require. Of course, it could be said that sometimes future happy experiences cause prior distress, and sometimes prior happiness. But quite how that could be explained — why sometimes one, and sometimes the other prior attitude is caused — I feel unable to anticipate.

Schlesinger's understanding of Ivan's distress is wrong because the idea of temporal movement is an absurdity. In the remaining part of this chapter I will try to show that the occurrence of the sort of feelings which are exemplified in the story of Ivan's concern can be adequately explained on a static view of time. This will be done by noting what it means to be a creature which aims to have its projects and schemes fulfilled and its desires satisfied (or as few desires as possible dissatisfied).

There are two possible conclusions which challenge Schlesinger's view of distress over unpleasant future experiences.

(1) There is no such thing as temporal movement, and since it is true that anxiety over unpleasant future experiences can be justified only if temporal movement is supposed, then no such anxiety can be justified. (The anxiety that we show in the face of expected unpleasant experiences would then be seen in its true light, having much the same character as, for instance, the anxiety that some primitive peoples have displayed when threatened with having their photographs taken, because their unwarranted belief is that the camera will capture their souls. Their fear is based on a mistaken metaphysical view, as would be, on this view, our fear of unpleasant future experiences.) This possible conclusion, which I shall not in fact defend, accepts the connection which Schlesinger is at pains to point out, between the legitimacy of anxiety over unpleasant future experiences and temporal movement, but since it rejects the possibility of temporal movement, it rejects also the possibility that anxiety over unpleasant future experiences can be justified. Schlesinger seems to hold that since such anxiety in fact occurs, then there must be some theory of time which explains it. This conclusion holds that if it is ever true that people do have anxiety over unpleasant future experiences, that is because they are deluded that there is temporal movement, or they believe that

such anxiety needs no justification. For this delusion to be legitimate delusion, we need to understand how the false belief in temporal movement *should* produce anxiety over unpleasant future experiences. This is fairly obvious. The erroneous idea of temporal movement has been modelled on spatial movement. So just as the speeding car heading directly towards us constitutes a threat which produces legitimate concern about the imminent impact, the unpleasant experience of visiting the dentist tomorrow is taken to be rushing at us in the same way such that essentially the same sort of anxiety is warranted. (And similarly for an expected pleasant experience.)

Someone who holds this conclusion would say that there are no facts about feelings towards future experiences which have to be squared with the static view of time. People who feel anxiety at the prospect of unpleasant future experiences have no reason to feel that way. If in fact there are people like this, or if indeed there are not, the static theory is unaffected.

(2) The second conclusion holds that Ivan's anxiety is justified, even though temporal movement is a myth. This conclusion rejects the premiss of the first conclusion, which says that anxiety over future unpleasant experiences is justified only if there is temporal movement. I feel we ought to favour this conclusion, because it takes on board both the fact that people *do* feel anxiety over unpleasant future experiences, and makes an account of why such anxiety occurs. If this conclusion can be supported, static theorists who sometimes feel as Ivan does are not displaying any inconsistency, indeed, there would be something odd about them if they didn't feel such anxiety in similar circumstances. Schlesinger's views on Ivan, temporal movement, and our feelings about the future, will have been refuted.

This conclusion can be supported by imagining Ivan saying the following. 'The crux of this issue lies in how we are to regard the fact that I (and other people of course) can anticipate what will happen later than the time of our anticipations. Given the fact of my arrest, interrogation, trial, sentence, and my knowledge of life in prison camps, I have good reason at t_1 to anticipate an unpleasant time during my period of imprisonment. This is so, because should certain circumstances prevail, then some of the projects into which I put my life will be frustrated, some irreparably so, and certain basic desires,[3] and certain non-basic but important desires will be frustrated. I believe that those circumstances *will* prevail, in which case I believe that certain of my projects and desires will be frustrated. The thought of those projects and desires being frustrated constitutes my current distress. I do not appreciate the likelihood of my projects and desires being frustrated *as well as* being distressed, as if I could have each without the other. The character of this appreciation *is* my distress. The prospect of circumstances ocurring later than my thought about it all, such that certain cherished desires are likely to be frustrated, including the general desires to avoid all pain, to be warm, well-fed, as well as specific projects such as that to finish writing the novel I am working on, provides sufficient reason for my current distress. This is consistent with the possibility that I may feel such distress even though later, whatever the circumstances, none of my desires or projects be in fact frustrated.'

The reason why Ivan's distress at t_1 is not alleviated by the nearby temporal presence of freedom and pleasant circumstances (immediately prior to t_1) is that such circumstances cannot contribute to the fulfilment of his present desires and projects, or desires and projects he can reasonably expect to have at times later than t_1. Our projects, and our basic desires, matter to us in a very simple way. When we see that a project of ours is actually being frustrated, or when we think that a project of ours will be frustrated, we feel varying amounts of distress, all according to how important the project in question is to us. Similarly for what I am calling our basic desires: when we feel pain, for instance, we also feel distress, and the thought that we are going to feel pain is itself characterised as distressing. (A masochist would, of course, find this a poor example, because pain for him, at least certain sorts of pain, bring the opposite of distress, pleasure. There is nothing logically necessary about the fact that we usually find that pain and distress go together. I take it that the masochist is experiencing the same sort of sensation when he reports pain, and enjoys it, as the sensation which I feel when I report pain but don't enjoy it. The masochist needs to be offered a different example, perhaps the basic desire not to feel uncomfortably thirsty would suffice.) That we are like this can be put down to evolution. A creature that feels distress when its projects are frustrated, thus taking action designed to fulfil projects as well as possible, is likely to succeed better than a creature who is not distressed by its projects being frustrated. By and large, the things which satisfy our basic desires are beneficial to the continuation of the individual, and therefore of the species, and the things which frustrate our basic desires are not beneficial.[4] As with the pain – distress connection, there is nothing necessary in this arrangement. It is logically possible that there be creatures who feel distress when they eat essential food; it is obvious why creatures so constituted are at a disadvantage.

(Our discussion will be helped if we can conflate the distinction between basic desires and projects on those occasions when both categories need to be referred to, or when making the distinction is immaterial. I will do this by using the term 'interests'. This term is appropriate because we know what it means to say that people like to have their interests satisfied, or when someone says that they are interested in achieving such and such. Using the term 'interest' sidesteps the important and difficult issues as to what we should say about whether it is in the subject's interest to have his compulsions or habits satisfied, or whether induced desires (induced by drugs for instance, such as addiction) should be viewed similarly to what I have called basic desires. Someone might try to satisfy his desire for more warmth, and someone might try to satisfy his desire for more heroin, but these interests are obviously very different in the sense that they seem to be different *types* of interest, not merely that they have different objects. I will avoid examples which make reference to unnatural or induced interests. Given my interest in the connection between the concept of time and what living life as a human being is like, that there is a problem here should not be a problem for me.)

We can see that Ivan's distress at t_1 results from the mismatch he sees between what he believes later circumstances will be like and the requirements of his interests. The fact that he believes some of his important projects will be frustrated, and that some, if not all, of his basic desires will be frustrated to various degrees, explains his distress and also provides sufficient reason for that distress to be legitimate. Of course, as some of us sometimes do, he may in fact appreciate the situation badly in a number of ways. He may anticipate correctly that certain circumstances will arise, but not appreciate that that being so, certain interests will be frustrated, or he might anticipate circumstances which do not in fact arise, even though he correctly appreciates that were they to arise certain interests would indeed be frustrated. He might have a weak perception of which of his projects really matter, and how they rank alongside each other, or of how badly he would be inconvenienced were certain basic desires frustrated, and make mistakes (that is, have inappropriate anxieties or hopes) as a result.

To accept that people do have such anticipations of circumstances later than the time of anticipating (never mind whether they are sound or not), we do not need to think that the events which are anticipated (or non-events should the anticipations be flawed) are moving through time, or have a real temporal tense.

Someone may object to this analysis saying: 'On your view, Ivan's distress at t_1 is generated by his realising that circumstances in the period t_1 to t_2 are likely to frustrate certain important projects which he has already commenced; those circumstances, if they turn out as anticipated, would mean that these projects will not further in the period t_1 to t_2. On top of that, Ivan believes that certain basic desires that he has (to be fed adequately, for instance) will also be frustrated. Yet, just as he can realise all that, he can similarly realise the analogous relation between past circumstances and his past interests. Why should he feel concern about his future interests being frustrated by future circumstances, but not feel concern about his past interests having been frustrated by past circumstances? It's no use saying that we *do* sometimes feel concern about our past interests having been frustrated, because that says too little, as we can see by imagining the following. We have to pay two unpleasant visits to the dentist. One visit took place yesterday, one will take place tomorrow. The degree of unpleasantness to be suffered tomorrow is gauged the same as the degree of unpleasantness experienced yesterday. It is obvious that our forthcoming visit causes much more anxiety than our past visit, yet our basic desire not to suffer pain will be frustrated to exactly the same extent *tomorrow*, as it was yesterday. Your account obviously leaves something out. Whereas Schlesinger's acceptance of temporal movement readily explains our differing anxieties towards the two visits to the dentist — the visit we are moving towards causes more anxiety than the visit we are moving away from.'

A weak reply would accept the apparent truth that the theory I favour cannot account for why it is that the frustration of future interests is of more concern to us than the frustration of past interests, but the transient theorist must also accept that equally the temporal movement theory cannot say why we should fear unpleasant events moving towards us more than unpleasant events moving away from us. The transient theorist claims that temporal movement supplies an explanation of our differing attitudes towards past and future; yet there seems no reason why we should not see this the other way round and say that our differing attitudes towards the past and future supply an explanation of why we are inclined to accept temporal movement.

Thankfully, there is a stronger reply which denies that the

analysis I favour cannot account for why it is that the frustration of future interests is of more concern to us than the frustration of past interests. If we consider the example of the two visits to the dentist, on the day before the second visit, which is also the day after the first visit, it is quite rightly claimed that we would be much more concerned about the forthcoming visit than the visit the previous day. At this time, the earlier visit is for us a memory, one which may well be remembered with its own fair share of unpleasantness. The later visit is something that we can only anticipate. We cannot at this time experience the actual unpleasantness of the visit, but we can, and usually do, experience the unpleasantness of the anticipation itself. On the one hand, we have a recollection, and on the other hand, an anticipation. The recollection is of an experience exactly similar to that which is the object of our anticipation. Yet the anticipation troubles us much more than the recollection; or if we feel we should resist the anticipation being compared directly against the recollection in this way, it seems undeniable that the anticipation concerns us in a completely different sort of way, even if it is true that the recollection which we happen to have is pretty unpleasant. Why should that be? The answer lies in taking a closer look at what it means to us to be creatures who live in time, who experience the events of our lives in succession, and who have projects which we strive to fulfil. It seems somewhat trivial to point out that past circumstances which frustrated past interests matter to us, if indeed they now matter at all, in a completely different way from that in which anticipated future circumstances matter to us, for the simple reason that there is often something that we can do now to influence how those future circumstances will turn out.

It looks to me as though this is why we find that our attention is concentrated upon the future in a way in which it is not concentrated on the past. Our expectations and anticipations attract our thoughts because what we decide to do now, and our acting on these decisions, usually makes a difference to how future circumstances will develop. Thus, acting in the present may well change an anticipation we have from one of an unwanted event that would have frustrated a current project in some detrimental way to one of an event that will not frustrate our project, but might even further it. An action we perform, whatever outward effects is has in the world, may, as well, operate directly upon the anticipations we have and relieve us of our anxiety about the future. For example, I can eliminate the anticipation (and the feeling of anxiety that having this anticipation involves), that I won't have a marrow to enter in the contest, by buying some marrow seeds and planting them in the appropriate way. Indeed, just to form the intention to go and buy some seeds and to plant them is all I need do to undermine my anxiety that I won't have a marrow to enter in the contest.

There is obviously a very real point to directing our attention upon our anticipations, and to be concerned about the future. Its being true (if it is) that past circumstances frustrated our past interests cannot concern us in the way that the present anticipation of unfavourable future circumstances grips our attention, because those past circumstances are beyond our reach, whereas the future circumstances may well be within our reach — that is, we can now, in principle, affect what happens, but we cannot affect past circumstances. (This will be queried later on when we consider the claims of fatalism.) This being the case is consistent with the fact that our anxiety is worse when we cannot see clearly the best way to influence future circumstances in our favour. In such a situation our experience is that of being arrested, to varying degrees depending upon the importance of our interests, by what may be called an urge, or a drive, or something like that, to find the right course of action; we are all familiar with what this sort of tension is like. Our anxiety is usually at its greatest extent when we can think of *nothing* to do that will influence future circumstances in our favour, because in our thoughts at the same time is the knowledge that *in principle, some* action or other, if only we could think of it (even if it's to get someone else to act), would save the day. Choosing from a range of equally unsatisfactory compromises similarly creates such tension. And our concern is at its least when we are confident that an easily attainable course of action will make circumstances favourable.

Our task is, in the present, to do what we can to make what happens satisfy our current interests and what we believe will be our future interests, such that our interests are frustrated as little as possible and furthered as much as possible. There is nothing that can be done now to make *past* circumstances match our past interests. Either they matched well, moderately, or poorly. The fact of the matter is of academic interest when contrasted with our interest as to what will happen next and to how the longer term future will turn out. If it is true that past circumtances poorly matched our past interests, this is a fact we know, assuming normal reliable memory. But that future circumstances match our future interests we obviously do not know. Merely being in a state of not knowing creates unsettling tension or uneasy apprehension.

What happened in the past can concern us only to the extent that knowing various facts can disturb us. But the power of the future to worry us is that much greater than the power of the past, because uncertainties (which to varying degrees is what the future is to us) are disturbing in their own way anyway, added to which comes the tension created by our striving to see how we can make future circumstances match our future interests.

The view that I am adopting here seems a little strained when we consider the case of someone who is facing an inevitable and unavoidable nasty fate, such as execution. I have suggested that we do not concern ourselves with the past, in the way we concern ourselves with the future, because we cannot act now to bring past circumstances into line with our past interests. The past is fixed, and there is little point dwelling on the point that, if it be true, certain past interests have been frustrated. If we look at the man awaiting execution, we can see that the event of his execution is altogether beyond his reach to influence, just as any of his past experiences are now beyond his reach. Since being concerned about the past is pointless from the point of view of satisfying current interests, because the past is fixed, it looks as though we ought to criticise the condemned man if we find that he is worrying about his forthcoming execution, because if anything is fixed, that event is. Yet, isn't he perfectly justified in worrying about his execution? It is hard to think what to say about this case, because there are facts which incline us in opposite directions. Firstly, it does seem intuitively correct to hold that a condemned man is perfectly justified in being anxious about his forthcoming execution. But the fact that there are cases on record where condemned men, recognising what their fate is to be and accepting its inevitability, appear to have given up absolutely worrying about their executions urge that our analysis is right after all — anxiety is appropriate only if it is worth the agent's while trying to work out which actions would help make circumstances satisfy his interests, or else frustrate them as little as possible. It could be urged that the condemned man who remains anxious about his execution has not properly realised the futility of his situation, that there are no actions of his which can make any difference to what happens. We might even try to explain the condemned man's anxiety, in a way which is consistent with our analysis, by suggesting that his is a different sort of anxiety, in that he *does* fully appreciate that there is nothing he can do which will make any difference to what happens, but that the anxiety he nevertheless feels is constituted by this very realisation. In general terms what the suggestion calls for is the possibility that a man recognise that he cannot act to influence a certain situation, and his seeing this makes him anxious. The sort of anxiety that I have been talking about up till now is that which a man has when he may well feel inclined to say 'What on earth can I do to get such and such interests satisfied?' The different sort of anxiety would be had by someone when they felt inclined to say 'What a shame I cannot act to influence such and such situation.' This to me seems after all to be the way we are inclined to worry about our past interests having been frustrated (if we are concerned about this at all) — 'What a shame I did not act [whether or not he is aware *how* he might have acted] to influence such and such (past) situation.'

This different sort of anxiety seems to me to be what we ordinarily call regret. Someone might regret that they acted thus, or failed to act thus, or that such and such happened or did not happen; and the condemned man can regret that nothing can be done to avert his execution. It seems appropriate to regret that the weather on Mars is bad (if I am trying to observe the planet's surface through a telescope), but that fact is not one I can be appropriately anxious about. If I were capable of affecting the weather on Mars, then it would be appropriate to be concerned about how I might change it, because feeling concern (amongst other things, such as feeling hope, or pleasant anticipation) is often part and parcel of what the experience of deliberating and working out courses of action is like for human beings.

I will finish this chapter by summarising my views on our differing attitudes towards past and future experiences. I believe the attitudes we have are perfectly comprehensible even if we reject Schlesinger's view that they can be comprehensible only if we conceive of events flowing towards the subject, and of the subject 'passing in and out of' periods of pleasant and unpleasant experiences. We are concerned about what we anticipate because what we do now will make a difference to what happens; if our interests are poorly met, it might be our fault. What we do now does not make a difference to what has happened, to whether our past interests were satisfied, or not, by past circumstances. It might help to utilise an idea of C. H. Whiteley's,[5] which he introduces in his views on intentional action, but which has application for our current discussion. Whiteley says that our awareness, of things in general, comes ready supplied with the attitudes of either assent or rejection. We either like what we are aware of, or we dislike what we are aware of. He says:

> All awareness is in the mode of acceptance-or-rejection . . . In being aware of something I accept it, approve of it, say Yes to it, or I recoil from it, repudiate it, say No to it. 'Nice' and 'nasty' are more fundamental and pervasive categories in experience than 'big' and 'small', 'dark' and 'light', 'swift' and 'slow'.

Our anticipations of what might befall us work in the same way. If we expect a certain event to occur, and we think its occurring will further our interests, our anticipation of the event will be in the assent-mode of awareness. If the event is believed to frustrate our interests, our anticipation will be in the reject-mode. Whiteley seems to have ignored the possibility that our awarenesses, and our anticipations, could come in a neutral mode. An anticipation in this mode would be of an event that is thought not to affect our interests for either better or worse. Talking like this is perhaps a complicated way of making the simple point that some things make us happy, and others, sad. I am happy when I think my interests will be furthered, sad if I think they will be frustrated. But the point I wish to stress is that we do not appreciate the likelihood of our interests being frustrated independently of feeling upset about it; our appreciation comes already switched on, in the case of our interests being frustrated, in the rejection-recoil mode. Because we can act only towards the future, our anticipations of what is to happen come in either the accept-mode (if we think our interests will be satisfied), or the anxiety-reject-mode depending on the size of the gap between what we think will happen and what we think we can do about it (if we think it likely that our interests will be frustrated), or the regret-mode where no action can be to any avail; whereas the anxiety-reject-mode, which I have earlier suggested goads us to action, is not a mode our recollections can come in, because we cannot act now to satisfy past interests — the appropriate modes for recollection appear to be regret, acceptance, or neutral.

Notes

1. See George N. Schlesinger, *Metaphysics*, Chapter 4.

2. *Metaphysics*, p. 109.

3. I have in mind here such desires as that to feel warm enough, not to feel thirsty or hungry, to feel loved, to feel valuable to those who matter. I believe we have an intuitive knowledge of what these basic desires are simply as a result of experiencing what it is like to live the life of a human being.

4. It is doubtful that we can say the same of all the things which satisfy and frustrate human projects, since sometimes people have projects which are satisfied only by their doing very dangerous things.

5. See C. H. Whiteley, *Mind in Action*, p. 52 ff.

19

'Thank Goodness That's Over'

Having talked about how we regard our future experiences, I shall now turn my attention to how we think of experiences we have just had. I am particularly interested in the fact that human beings feel a sense of relief when a nasty experience has just finished happening. One way we express this relief is to say (in a relieved sort of way) 'Thank goodness that's over!' meaning the 'that' to refer to the past nasty experience which has just finished. It would be perfectly appropriate for Ivan, when released from the prison camp at t_2, to exclaim 'Thank goodness that's over!' intending to refer to the whole of his prison sentence.

A. N. Prior suggests[1] that the fact that we use this expression, and variations on it as well of course, to express our relief upon the ending of our suffering some unpleasant experience, implies that events really are past, present, and future, over and above their being later than or earlier than, or simultaneous with, other events. To believe this is also to believe that the tensed theory of time is true. Since I believe that tensed facts are a myth (the unpleasant experience referred to in 'Thank goodness that's over' isn't *really* past, although it is earlier than some other events and later than still others) I am compelled to adopt one of two views; either the expression 'Thank goodness that's over' is in some sense illegitimate and always inappropriate, the exclaiming of which is beyond rational comprehension, or else 'Thank goodness that's over' makes sense without requiring that that understanding make reference to tensed facts. The latter view, I believe, can be substatiated.

Let us take the simple example of someone, S, having a headache which endures from t_1 to t_2 (three hours, say). We can imagine S saying just later than t_2, meaning to refer to the headache, 'Thank goodness that's over.' This statement, or exclamation, or whatever it is, would appear to have truth conditions, since it contains the factual phrase 'That's over' which is either true or false so long as S intends it as a genuine statement of fact such that the 'that' refers to some earlier event. So in the example of the headache, 'That's over' is obviously short for 'The headache's over', and that will be true if and only if the date of its utterance is just later than the time at which the headache ceases (t_2). Thus *if* 'That's over' is true, it is necessarily the case that the date of the statement is just later than the date at which the headache stops (usually very soon afterwards). But its being true that this token of 'That's over' occurs just later than t_2 is timelessly, or tenselessly, true: ' ''That's over'' occurs just later than t_2', if such be true, is a tenseless truth, true at all times. It is also tenselessly true both that the headache ceases at t_2, and that the ceasing of the headache is just prior to the utterance 'That's over.' Why should S want to thank goodness for any of these tenselessly true facts? And why should he thank goodness at the time he does rather than any other time? The tenselessly true facts that 'That's over' occurs just later than t_2, that the headache ceases at t_2 and that the former occurs immediately prior to the latter, can be thanked for at any time S wishes. Prior's view is that S can't really want to thank goodness for these facts, but means to thank goodness for the fact that the headache is not a present experience, but is now in the past. So, when S (or anyone) uses the expression 'Thank goodness that's over' he must be thanking goodness for the tensed fact that some unpleasant experience is past; if S's headache doesn't really have the temporal property of presentness, which it loses to acquire the temporal property of pastness, S wouldn't have anything to thank goodness for.

Prior takes a tenseless account of this matter to fail, if I may reiterate his point, because the tenseless fact which makes 'That's over' true, namely the fact that 'That's over' occurs just later than the headache, always was true and always will be true. If S is thanking goodness for *this* he can do so at any time he wishes. The fact that 'Thank goodness that's over' is only ever used at the conclusion to an unpleasant experience indicates that something else is being thanked for — the fact that the unpleasant experience is past. That is a tensed fact, expressed by an A-statement which, we may recall from Chapter 6, places an event at a temporal location some distance or other from the present. And an event's having such a location is over and above its having B-relations with other events (its being later than some events and earlier than others). The tenseless fact that a headache of ours and a statement of ours are in a certain B-relation just doesn't seem a thing anyone would want to thank goodness for. The only other candidate that we could possibly be thanking goodness for is the tensed fact that the headache is past.

There are at least three difficulties with Prior's view, even if for the moment we ignore the convincing reasons put forward in Part I which indicate that temporal properties are a myth, and that there cannot be tensed facts. That a nasty experience of mine is past is not sufficient for my exclamation 'Thank goodness that's over' to be appropriate. I cannot use the phrase 'Thank goodness that's over', referring to a visit paid to the dentist a year ago, to mean what I mean if I say the same about a visit just this minute concluded. The relief that 'Thank goodness that's over' usually expresses is not appropriate in a context where I am mentioning an unpleasant experience that happened to me quite some time ago. The fact of our experience is that we regard recently past experiences differently from the way we regard experiences that are more past. Yet, if I am really thanking goodness for the tensed fact that a nasty experience is past, it is presumably just as legitimate for me to thank goodness for the pastness of the visit to the dentist last year, or even the visit that occurred ten years ago, as it is for me to thank goodness that yesterday's visit is past. But this is not how we really experience our lives. If said with strong feelings of relief, 'Thank goodness that's over', referring to an experience years ago, would be taken as a sort of joke, or as a way of demonstrating the truth of what I just said, that indeed, we feel very differently about recent experiences from the way we feel about those that happened longer ago. In response, Prior could make an *ad hoc* amendment to his theory and stipulate that 'Thank goodness that's over' is appropriate only when the experience referred to is suitably recent. But this stipulation doesn't seem adequately accounted for by just pointing to the truth (as Prior takes it) that earlier events really are *past*. If it is not appropriate for us to say 'Thank goodness that's over' of events that are more past than recently past, we can call for an explanation of why that should be so. What is different about events more past other than that they are more past? I cannot imagine how an answer to this might go, in which case Prior, to my mind, is struck with the stipulation that 'Thank goodness that's over' is appropriate only if said of recent unpleasant experiences, which cannot receive an explanation in terms of events being tensed, or of there being tensed facts.

The second difficulty which Prior's theory faces is what I believe is its inability to account for the following unusual but possible turn of circumstance. S's headache lasts from t_1 to t_2; but at t_2, just as the headache is noticeably declining, and all things being equal S's exclamation of 'Thank goodness that's over' would be entirely appropriate, S feels a pain in his foot (say) which quickly assumes the level and degree of discomfort which the headache had. In this situation S would not utter 'Thank goodness that's over.' There is something very unsatisfactory in trying to picture the legitimacy of his uttering, with feelings of relief, the statement 'Thank goodness that's over' whilst at the same time he is grimacing with the pain in his foot. Yet the headache is past — surely a fact that S would feel like thanking goodness for? Or even if he doesn't feel like thanking goodness, it is, isn't it, a fact he can now be thankful about? Of course, it will be said that the pain in his foot diverts his attention from the fact that his headache is now past. But even in the case where this does not occur, and I see no reason to hold that it must occur in every instance (that is, S could be aware that the headache has ceased, even though he is also aware of the new pain in his foot), it would not be appropriate for S to thank goodness that the headache is past, even though, on Prior's view, the headache is in fact past. (This example offers the clue that 'Thank goodness that's over' being appropriate is partly determined by what the subject's experiences are like at the date when uttering the exclamation would ordinarily be expected, a fact we will return to later.) Again, Prior could stipulate that a further unpleasant experience must not be present to the subject at the time of his saying (with respect to a suitably recent unpleasant experience) 'Thank goodness that's over' — the utterance of which would otherwise be appropriate. But this stipulation too appears to find no account in viewing our experiences as tensed.

The third difficulty for Prior's theory is its inability to explain another inappropriate use of 'Thank goodness that's over.' Here, we imagine that S has had a steady and enduring pain for quite a long time, say two months, but then the pain steadily declines until after a further two months, say, it has gone. In this situation, S would not feel inclined to say 'Thank goodness that's over' at any stage. Even though it is undeniably true that at the end of the whole four month period S's unpleasant experience of the enduring pain is past, and suitably recent as well, *and* his attention to this fact is not diverted by the appearance of another pain, he would not cry with relief 'Thank goodness that's over.' Obviously he can be thankful in a sense that he is not now in pain, but this sense seems to be that in which S can be glad that any fairly substantially nasty experience is past. In this situation there is no relief that 'Thank goodness that's over' can express, in contrast to the situation in which the pain ceases quite abruptly and 'Thank goodness that's over' clearly would express something quite different. A tensed view of these situations is not able to bring out these differences or explain why we should experience them like this.

The general difficulty with Prior's tensed view which these examples expose is that in saying merely that nasty experiences which we have had really *are* past, carrying the full implications of what the tensed view of time maintains, not enough is said to provide an explanation as to why 'Thank goodness that's over' is an appropriate thing to say in a fairly limited range of cases. And since temporal properties cannot be had by our experiences anyway, it is not a question of finding ways to reconstruct the tensed view in the hope of generating a more satisfactory picture.

The problem with Prior's ideas on this, and the difficulties which Mellor gets into about what 'Thank goodness that's over' is doing, are resolved by my view of this matter, which I will go into after looking at what Mellor says on this.

D. H. Mellor[2] rightly queries Prior's view that we are thanking goodness for facts (namely tensed facts) when we say 'Thank goodness that's over.' Instances of 'Thank goodness' he believes express a feeling of relief. The feeling of relief naturally occurs at the conclusion of an unpleasant experience, just at the time when the statement 'That's over', referring to the experience, is true. Accordingly the two statements get uttered at the same time, giving the impression that we are thanking goodness *for* something, when in fact we are only expressing the feeling of relief. If we are not thanking goodness for anything, then even if we thought there could be tensed facts, such tensed facts would have no role to play in any view about what implication our uttering sentences with the import of 'Thank goodness that's over' will have for our metaphysical beliefs.

One reason why Mellor thinks that 'Thank goodness that's over' does not really thank goodness for anything, but that its function is to express a feeling of relief, is because we can disassemble its component phrases, switch them round and get 'That's over: thank goodness.' Mellor feels that this makes it much easier to accept that the phrase 'Thank goodness' in fact is nothing more than the expression of relief.

This reason I think is unconvincing. The fact that we can thank goodness for someone's arriving at last (say) with the expression 'He's come: thank goodness' (as opposed to putting it the other way round and saying 'Thank goodness he's come') would obviously not be taken as conclusive proof that we are not really thanking goodness for something. That we can in fact switch around the phrases and retain our meaning when we use expressions like 'Thank goodness that's over', if it is relevant to an analysis, what relevance it has is not nearly as obvious as Mellor appears to think.

The idea that 'Thank goodness that's over', when uttered in the sort of context we are interested in, expresses a feeling of relief, is plainly correct. If this needs to be shown, we can do it by pointing out that in a situation where someone has just ceased to suffer in whatever way they were suffering, and they utter a long sigh in a relieved sort of way, we have to accept that the sigh (in the circumstances) means just what 'Thank goodness that's over' would have meant if it had been uttered instead of the sigh. Clearly, Mellor is not wrong to talk about relief. It's what he says about the relief which misses something important for finding the right view on this matter. Mellor says, 'What a token of "Thank goodness" really does is express a feeling of relief (not necessarily relief from or about anything, just relief).'[3]

The sort of relief that Mellor wants is relief that doesn't have an object, to be compared with happiness, which is not happiness *about* anything, but is just a feeling of well-being. I don't believe that there can be relief of this sort, because such relief and a general feeling of well-being would not be distinguishable. This can be shown by noting that it is logically possible that object-less feelings like nausea, well-being, dizziness, and others be caused by drugs: indeed many such feelings can in fact be so caused. Let us imagine there to be two drugs, one which causes well-being, and the other which causes the sort of relief which Mellor requires. We are given one of the drugs, but we don't know which one. It begins to take effect, so we are obviously feeling either the sort of relief that Mellor requires, or general well-being. It seems plain to me that we would not know which of the two feelings we were in fact having, yet that outcome is patently absurd; if two feelings really are distinct, and we know we had one of them, we would also know *which* one we were having. Relief is obviously a type of happiness, but if we are not aware that that feeling is directed at anything, we would not be having a feeling that is phenomenologically distinguishable from general well-being. It is, of course, perfectly conceivable that in some circumstances what would ordinarily give rise to relief in fact produces a feeling of well-being in the subject. But that has no real bearing on the fact that in our experience we can tell the difference between relief and well-being, and when we feel relief we know what the feeling of relief is about. Mellor is wrong when he says that we can have feelings of relief that are not necessarily relief from or about anything.

Mellor is worried, I think, that we might have to model relief on belief and other propositional attitudes, and if we did have to, just as the object of a belief is a fact, the object of a *relief* would be a fact as well. Mellor would argue that that must be wrong, because the only sort of temporal facts he can admit are tenseless facts, and feeling relief cannot be a matter of feeling relieved that tenseless facts are true, as we have seen. The feeling of relief certainly cannot be relief that a tensed fact be true, because, says Mellor, there are no tensed facts. So the option Mellor takes is to say that relief has no object, which I have just argued cannot be right either.

The answer to this confusion is to show how a feeling of relief can have an object, but in a way that is different and somewhat more involved from the way propositional attitudes, such as belief, have objects. This view, which I shall now explain, does not seem to offer a positive argument for the truth of the static view of time, but I think it shows that the way in which we feel relief is consistent with time being static. In other words, that time is static does not seem to be necessary for the intelligibility and plausibility of my view. On the other hand, the way I think pain and relief are related will not require us to conceive particular pains and reliefs as being tensed.

Not only do human beings experience what happens to them in succession, we also experience the succession itself. That is, if S experiences B (whatever the experience is) after A, S also experiences B happening after A.[4] The circumstances in which relief arises are those where S experiences something nasty and then something nice, or at the very least something neutral or not-nasty. It is not enough that S experience something nasty, and next he experience something not-nasty. If, at the time he becomes aware that he is experiencing something not-nasty he is no longer aware that something nasty has just been experienced he could not feel relief (but he might feel happy or contented I suppose). My understanding of this is that relief arises only when S is aware of the *contrast* between what he is currently experiencing and what he has just been experiencing. Such contrast between differing kinds of experience (and our interest here is in comparing nasty with nice or not-nasty experiences) seems to be experienced itself as a whole package, so to speak. If we were not directly aware of 'event packages' in this way, we would not be able to distinguish between hearing a chord and hearing the notes of the same chord played as an arpeggio. If S is aware of his headache ending, he must be aware both of the current lack of headache, and his previously suffering the pain of the headache: he must be aware of the contrast between the two experiences. Obviously, the headache must have happened, and the state of painlessness must be going on for S to be aware of this contrast. Notice also that in a few hours time, or the next day, he will no longer be aware of the contrast itself, although he may still recollect it as a fact that the headache ceased at such and such time, which was when the painlessness commenced. But now, the ceasing of the headache is something he cannot experience, because at this later time he cannot be aware of the contrast between pain and painlessness. Our awareness can occur only at the time when the events of which we are aware happen.

Let us recall what Ivan said about his distress that his interests would be frustrated. He said that he did not appreciate the likelihood that his interests would be frustrated *as well* as being distressed; the character of his appreciation *is* his distress. This is a particular instance of Whiteley's general point (referred to at the end of the previous chapter) that our awareness of things occurs in either the assent-mode (we like what we are aware of) or the reject-mode (we do not like what we are aware of). (We saw before the need for a neutral-mode to cover those circumstances where we don't much mind the occurrence of what we are aware of.) It seems to me that relief is in fact a particular 'mode' of awareness — to say just that S's awareness of current painlessness and immediately prior pain as a package is necessary and sufficient for his being relieved that his headache has ceased, although correct, misses the point I want to make about this. And that is that relief is a mode of awareness, in the sense that Whiteley thinks awareness can have modes. The simple point is that the character of our awareness of the sorts of contrast in the experiences we have been talking about is just relief.

A false view of this would be to hold that we have two distinct capacities, one that of being aware of the contrast between a recent headache and current painlessness, and the other that of feeling relief about it. Relief is the mode in which our awareness of such contrast in our experiences comes.

This view of relief I believe meets the difficulties discussed earlier. Firstly, we can now say in what sense relief has an object. If we are aware we are necessarily aware *of* something. If we say that awareness has an object, that is what we mean — the object of my current awareness is whatever it is I am aware *of*. Accordingly, the sort of awareness we have been discussing has as its object a contrast between a nasty and a not-nasty experience. Since relief is the mode in which this awareness comes, there is a clear sense in which relief can have an object, which thankfully accords with our everyday experience, in that the question 'What are you relieved about?' is obviously intelligible and usually solicits a meaningful answer. The answer we are likely to get makes reference not to the contrast between nasty and not-nasty experiences, but to the nasty experience alone. But I hope I have shown that that reference wouldn't make any sense unless we understood the subject to be aware of the contrast between the experience he mentions and what he is now experiencing. Being aware of the ways in which our experiences are changing is not being aware of the facts; if I am aware that I have a headache, I am not aware of the fact that I have a headache, I am aware of the headache itself — and being so aware entitles me to make the factual claim, if I so wish, that I have a headache. This interpretation overcomes Mellor's concern that the object of a relief should not be a fact. Relief has an object, but in a somewhat involved sense, in that if I am relieved that my headache is over (perhaps indicating this by saying 'Thank goodness that's over'), the steps we can take to 'explain' this, and so say to what the relief attaches, are limited to pointing out that this relief can be appropriate just so far as these conditions are met: I have had a headache, which has recently ended and I am now having a contrasting not-nasty experience; I am aware of this package of experiences, specifically I am aware of the contrast between them; that awareness comes in the relief-mode.

Murray MacBeath, whose thoughts are endorsed by Mellor's later thoughts,[5] offers an account of this issue which captures part of the account I have just rendered. MacBeath would choose to unpack my notion of being aware of the contrast between a painful experience and a non-painful experience immediately subsequent to it in terms of *belief*. That is, at the time the non-painful experience begins, S must remember that he has indeed just been in pain (if he does not so remember, there will be no contrast for him to appreciate), and for him to do this he must have the tensed belief (a belief whose articulation would be in temporally tensed terms) that the painful experience is past. I have no serious objections to elucidating the notion of 'being aware' in terms of memory and belief: to be aware that something was recently the case (for oneself) implies that one has a memory of what was the case, and having a memory implies that one believes that something or other, in this instance, was the case. However, with respect to the headache example (which MacBeath hardly addresses directly because he deals with a different example of his own) it seems odd to talk in terms of having the *belief* that a headache has just finished, when we would usually talk in terms of *knowing* (for oneself) that this is the case. Though this is perhaps strained as well, since if someone claims to know something, we should allow that they could be mistaken, and it seems wrong that in ordinary circumstances someone could be mistaken as to whether or not a headache they have had has just ceased. This is perhaps beside the point, since if S has the sort of belief about his recently ceased headache that MacBeath thinks he has (that is, the belief that the headache is indeed recently past) we can see that the appropriateness of this belief rests on the tenseless condition that the belief occur after the headache stops.

MacBeath's account appears to miss out something. The problem here is that S's belief that a certain headache was had in the past between the past times t_1 and t_2, can be had by S any time after t_2, and MacBeath seems not to have an account of why relief arises when this belief is had just after t_2, and not at various times thereafter. If we accept that S can experience an 'event package' of pain then no-pain, being aware in the 'relief-mode' at the time the non-painful experience begins of the contrast between pain then non-pain, the problem is solved. My belief is that we all know what it is to experience an 'event package' since, for example, not only do we experience a succession of notes, we also hear the complete arpeggio. Someone who hears *merely* a succession of notes would not hear the arpeggio, just as someone who experienced *merely* a succession of headache and no-headache would not feel relief.

Anomalies pointed to earlier can now be cleared up.

The reason why I cannot now feel relief that an unpleasant visit to the dentist a year ago is over, is because the time is long past at which I can be aware of the contrast between the painful experience of my tooth being drilled and the painlessness that followed. I may of course be glad that my tooth is not at this moment being drilled, but that is neither here nor there with respect to my being aware or not being aware that my experiences contrast in the way we have discussed. It seems plain to me that the awareness which my relief really consists in can occur only at the time when the not-nasty experience, which is contrasted in the awareness with the immediately prior nasty experience, has started to happen. This is exactly similar to the awareness we have of the succession of events generally. I can only be aware of so-and-so's knocking followed by his entrance just after his entrance has occurred. Later than that time we would have to say that my reporting that events went this way results from my now remembering what I had earlier experienced. And what I had earlier experienced was the package of events comprised by the succession of the knocking and the entering.

So now we can explain why S cannot say 'Thank goodness that's over' sincerely if he refers to a headache just ended, if his foot is hurting him. Simply, there is no contrast between his experiences of which he can be aware in the relief-mode. He is certainly aware, we may suppose, of the succession of pain in the head and pain in the foot, but the awareness of a package like that does not come in the relief-mode, but, most usually, in the reject-mode.

Similarly, if S has a pain which gradually subsides, we can see that at no point does S experience a contrast between painfulness and painlessness. Accordingly there is no awareness of the sort of contrast that he would need to have were we to ascribe relief to him. Of course, he can be aware that his pain is subsiding, and that awareness, unless he is a masochist, will come in the assent-mode.

There is a general consensus with respect to the modes in which our differing awareness comes. Most of us are relieved when a painful experience gives way to a painless one; most of us are distressed if we understand that our interests will be frustrated in the near future; and most of us do not feel relief if one pain is replaced by another. The view I have urged at no point calls for a tensed view of time. It seems perfectly satisfactory to me to say that we can be aware of experiences that have just stopped happening to us if we conceive of such experiences standing in B-relations with our later awareness. If there is any compulsion to view our experiences as tensed, I am quite oblivious to it. To say that we are relieved, if we are, only about past unpleasant experiences, seems no more helpful than saying that we are relieved about unpleasant experiences that are earlier than our relief. In so far as I have already given reasons why I think tensed facts are a myth, and the overall dynamic view of time a nonsense, this current analysis of what is going on when we express our relief, with or without saying 'Thank goodness that's over', does not in the least incline me to embrace tensed facts.

Lastly, it might be asked 'Why does our awareness come in *these* particular modes? Why do we feel *relief* when painfulness ends and painlessness commences? Our awareness that that is happening to us us could come in any mode whatever.' The only answer I have is a simple one, and that is that the modes we happen to have are the best ones for the evolutionary success of the species. I hinted at this earlier. The way we feel about what happens to us seems to be cleverly worked out by nature to make us avoid danger, move towards what is good for us (in terms of survival, procreation), and to plan against future contingencies.

Notes

1. A. N. Prior, 'Thank Goodness That's Over'.
2. D. H. Mellor, ' "Thank Goodness That's Over" '. See also *Real Time*, pp. 48 ff.
3. ' "Thank Goodness That's Over" ', p. 24.
4. This is not a necessary truth. Sometimes, people with some sorts of brain damage, people who are senile, and others, are not aware that their experiences are successive, although obviously the experiences they do have occur in succession.
5. MacBeath, 'Mellor's Emeritus Headache', and Mellor, 'MacBeath's Soluble Aspirin'.

20
Fate

One of the most striking features about the sort of lives nature has bestowed upon us is our ability to influence a certain number of events to go rather as we wish, rather than as we do not wish. The influence we have, when we are honest to ourselves about this, is very partial. For a start, no past event is within our power to influence, and of all those events that might realistically be supposed to happen in the next few minutes, not very many of them are within my power to bring about, or not bring about, as I choose. Fortunately, some are. Whether or not I drink my coffee now, or let it go cold, is up to me. If I am on a journey, whether I turn left or whether I turn right at a particular junction, is up to me. Most people don't give this fact about our experience very much thought: we have grown accustomed to the way, and to what extent, we can influence events, that, mostly, we have no call to think about this fundamental feature of human life. Obviously we think about the choices which face us — what to spend our money on, which books to read, and so on — but we usually do not think in the abstract about the fact that genuine choices do await our decisions. We get on with our choices and live our lives, very rarely confirming to ourselves that yes, indeed, it *is* up to me what happens next.

The way we speak about choosing (when we get down to thinking out the metaphysics of this feature of our lives) urges us to accept the reality of the 'open' future. We choose from a 'range of possibilities'. The possible event of my drinking my coffee before it goes cold seems to lack something which it gains the moment I *decide* to drink it now and not later. With respect to the coffee, there are at least two possible future events, one being the coffee's steadily cooling, the other my drinking it. (There are others, of course. The coffee might get spilled; it might, since it is logically possible, shoot up into the sky and never be seen again.) Out of all the possible future events there are, when I decide to do something, and strive to bring it about, what I do appears to invest one or a number of these possibilities with, first, virtual certainty, and then, as they occur, actuality. An event, in becoming present, ceases to be a mere possibility, and becomes actual.

The way we experience our lives urges us two ways. One way, as I have just indicated, appeals to the 'open' future, carrying with it the implication that since future possible events 'become' actual when they achieve 'nowness', time must be dynamic. Since there is this difference between what we refer to as future events, present events, and past events — future events are as yet 'possible' whereas present and past events are in some sense 'actual' or 'real' — then there must be a real difference between past, present, and future events. In other words, all events, including those we believe we bring about, are really tensed. This is what the static view of time denies. On the face of it, either the belief that sometimes what happens is up to us must go, or the view that time is static must go.

For many people, experience can sometimes turn us the other way. Sometimes it can seem that really things are not at all up to us. Freedom is an illusion; we find that nothing we aim for results in success, our actions fail to steer us clear of what we all along had hoped to avoid. It can seem more and more obvious that we are fated to experience just one definite future which time has stored up for us, and that really there are no possibilities, there is no 'open' future. This view is consistent with both the static, tenseless view of time, and with the tensed view of time. Someone who holds that in reality the future is fixed and that there really are no future possibilities, appears able to do so without holding as well that events really are tensed. And, on the other hand, it seems equally consistent for a believer in the fixed future to hold that events *are* tensed, it's just that given that an event has a future tense, it might be argued, it does not follow that its happening or not happening might be up to us. Indeed, that an event has a future tense implies that when the time comes it has to happen. Whatever *we* want and strive for cannot make any difference to its eventual happening.

This second way of thinking, that things are not at all up to us, finds philosophical backing in the doctrine of fatalism. I propose now to state what the fatalist's thesis is. I will then in subsequent chapters explain why this thesis is mistaken, aiming to demonstrate that a 'non-fatalist' position does not require that events be tensed, and does not require that the future really be 'open', a 'realm of possibilities'. That is, it sometimes is up to us what happens, and it is sensible for us to aim for particular goals and to try to avoid events we would rather not experience, *and* it is the case that time is static. For ease of expression, I will refer to this view as the 'libertarian' position. The libertarian holds a view contrary to that which the fatalist holds: the fatalist holds that *all* events are unavoidable, the libertarian holds that *some* events are avoidable.

There are two formulations, each of which aims to establish the fatalist's conclusion that everything happens unavoidably in consequence of which human action is never to any avail. Both views will be examined in later sections. This what the views say:

(1) I will follow Thalberg[1] and call the first sort of fatalism 'truth-fatalism' for reasons that will be clear in a moment. It is also that formulation of fatalism which philosophy has toyed with from the very beginning. The truth-fatalist argues like this: the mere fact that a certain event, E, is going to happen guarantees that what will occur (in this case, E) cannot be avoided. Generalising, we derive the conclusion that whatever happens is unavoidable. Whatever will be will be, says the fatalist, regardless of what we do, in which case there is never any point in our trying to avoid what is destined to happen. Consequently, deliberating over what we should do is a waste of effort, and trying to achieve something or other by taking a certain course of action will never make anything happen that was not all along fated to happen. 'So nothing that does occur could have been helped and nothing that has not actually been done could have been done.'[2]

The metaphysical truth of the matter is that there is a parity between the past and the future. As Richard Taylor points out, a fatalist regards what is *going to* happen in the same way that everyone regards what *has* happened. The past is fixed and unalterable; nothing can be done to alter the occurrence of a past event. We are all fatalists in the way we look back on things, says Taylor. The fatalist objects to the way in which the future is popularly conceived, as being somehow obscure and somehow a 'realm of possibilities'. The fatalist is honest enough to reject these notions, which arise solely from our ignorance about what, as a matter of fact, will happen. Our control over past events is the same as our control over future events — that is, we are powerless with respect to both kinds. 'We say of past things that they are no longer within our power. The fatalist says they never were.'[3]

Taylor in fact argues for the fatalist position. At the centre of his exposition is the concept of 'body of truth'. He says:

> . . . it seems natural to suppose that there is a body of truth concerning what the future holds, just as there is a body of truth concerning what is contained in the past, whether or not it is known to any man or even to God, and hence, that everything asserted in that body of truth will assuredly happen, in the fullness of time, precisely as it is described therein.[4]

Taylor clarifies this notion by asking us to suppose, just for the sake of argument, that God exists and is omniscient, thereby knowing everything that is true. It follows that God knows all those things that are true of a particular person, Osmo. He knows when Osmo was born, and in what circumstances, and He knows when and how he will die. And He knows what happens to Osmo at each and every moment of Osmo's life. This body of truth about Osmo, we are now required to suppose, is revealed to a chosen prophet, who we may also suppose has no other source of information about Osmo, and who indeed may well have lived a long time before Osmo was born. The information which the prophet receives from God is written in a book called *The Life of Osmo, as Given By God*. The book is published, and distributed to libraries.

When a young man, Osmo chances upon a copy of this book, and is intrigued to discover that the book is about him. There can be no doubting the identity of the person the book is about, for the book begins: 'Osmo is born in Mercy Hospital in Auburn, Indiana, on June 6, 1942, of Finnish parentage, and after nearly losing his life from an attack of pneumonia at the age of five, he is enrolled in the St James School there.' And so the book goes on, in the 'journalistic' present tense, detailing all the main events of Osmo's life under the chapter headings 'Osmo's Tenth Year', 'Osmo's Eleventh Year', etc. Osmo's intrigue turns to amazement when he discovers that not only does the book contain facts about his life up to the present, but tells also of how he found the book itself, of the circumstances of his reading it, and facts about the future portion of his life which he has yet to live. Osmo is disturbed to find that, according to this mysterious book at least, he has only a few more years to live, since the final chapter is 'Osmo's Twenty-Ninth Year', which concludes with the words: 'And

Osmo, having taken Northwest flight 569 from O'Hare, perishes when the aircraft crashes on the runway at Fort Wayne, with considerable loss of life . . .' Osmo, reasonably he thinks, resolves never to get on that plane to Fort Wayne. Taylor concludes his Story of Osmo thus:

> (About three years later our hero, having boarded a flight for St Paul, went berserk when the pilot announced they were going to land at Fort Wayne instead. According to one of the stewardesses, he tried to hijack the aircraft and divert it to another airfield. The Civil Aeronautics Board cited the resulting disruptions as contributing to the crash that followed as the plane tried to land.)[5]

Taylor remarks that God, the prophet, and the book, were mere devices, introduced for illustrative purposes only. With regard to each and every one of us there is a body of truth made up of all those statements that are true of us, and excluding all those that are false. This is the case whether or not that body of truth is recorded (as it was in Osmo's case) and whether or not anyone happens to become acquainted with any of it. We all should believe what Osmo came to believe at the very end, that whatever we do, that whatever happens to us, is unavoidable, since whatever *does* happen must accord with the body of truth about our lives; whatever that body of truth says is going to happen must of logical necessity happen to us. Taylor remarks:

> Each of us has but one possible past, described by that totality of statements about us in the past tense, each of which happens to be true. No one ever thinks of rearranging things there; it is simply accepted as given. But so also, each of us has but one possible future, described by that totality of statements about oneself in the future tense, each of which happens to be true.[6]

Osmo discovered that he was powerless to render false that statement contained in the body of truth about his life which said that he would die in the aeroplane crash. He was of course similarly powerless to render false any other statement contained in that body of truth. We are all in Osmo's shoes. We are no less the victims of our own fate than he was of this. When we do something and achieve the desired result, that is because the body of truth about our lives contains the fact that that is what, at that time, we would achieve. Self-congratulation or praise upon such a success would be logically inappropriate, since had we aimed at some alternative, we would have failed. That I get what I aim for is no indication that I am really free to make happen what I want to make happen, and that really I could have done something else if I had wanted; it indicates no more than the fact that all along it was my fate to have that, and not anything else, happen to me. And in any case, what I deliberate about and the decisions I reach are fated to occur just as they do, just like everything else that happens.

(2) The second formulation of fatalism hinges on the concepts of sufficiency, and its companion, necessity. (I am going to call it 'condition-fatalism', because of its reliance on the connection between events and the conditions that must obtain for those events to happen.) It is argued for by, again, Richard Taylor.[7] We are to imagine that Taylor is a naval commander about to issue his order of the day. We are to assume that 'within the totality of other conditions prevailing, [his] issuing . . . a certain kind of order will ensure that a naval battle will occur tomorrow, whereas if [he issues] another kind of order, this will ensure that no naval battle occurs'.[8] Taylor takes 'ensures' to be synonymous with the usual meaning of 'is sufficient for', the standard explication being: '. . . if one state of affairs *ensures* without logically entailing the occurrence of another, then the former cannot occur without the latter occurring. Ingestion of cyanide, for instance, *ensures* death under certain familiar circumstances, though the two states of affairs are not logically related.'[9] We will also need to know that Taylor takes 'is essential to' to be synonymous with 'is necessary for', his definition being: '. . . if one state of affairs is *essential* for another, then the latter cannot occur without it. Oxygen, for instance, is *essential* to (though it does not by itself ensure) the maintenance of human life . . .'[10] Consequently:

> . . . if one condition or set of conditions is sufficient for (ensures) another, then that other is necessary (essential) for it, and conversely, if one condition or set of conditions is necessary (essential) for another, then that other is sufficient for (ensures) it.[11]

There are two further presuppositions which are required for Taylor's argument. One is that no agent can perform a particular act if there is lacking, at the time of acting or any other time, any necessary condition for the performance of the act. (Taylor offers as examples the fact that the presence of oxygen is necessary for his continued living, having learnt Russian is necessary to his reading a page of Cyrillic print, and having been nominated is necessary for his winning a certain election.) The other presupposition required is that 'any proposition whatever is either true or, if not true, then false'.[12]

Taylor is about to order either (1) that there be a naval battle tomorrow, this order ensuring that tomorrow a battle will take place, or (2) that there be no naval battle tomorrow, this order ensuring that tomorrow no naval battle will take place. Is it up to Taylor which sort of order he issues? He believes it is not. For, either it is true that a naval battle ensues tomorrow, or it is true that a naval battle does *not* ensue tomorrow. The first alternative we can refer to as B (standing for battle ensues) and the second alternative we can refer to as NB (standing for no battle ensues). If B is the case there is lacking a condition essential for Taylor's ordering that there be no battle (since NB being the case is essential for his ordering that there be no battle); and if NB is the case, there is lacking a condition essential for Taylor's ordering that there be a battle (since B being the case is essential for his ordering that there be a battle). Either B or NB is the case. It follows that it is not up to Taylor which order he gives. He can give an order, but he cannot choose which one that is to be. The order he does give must be in accordance with how events turn out tomorrow. Unless those states of affairs essential to his ordering that there be a battle obtain, it is logically impossible for Taylor to order that there be a battle (and similarly for his ordering that there be no battle).

This argument is supposed to be completely general. Whatever we try to do can be done only if those circumstances (past, present, and future) essential to our act obtain. The reason we perform any act at all is because we believe that it will ensure an outcome which we favour. What Taylor's argument establishes, if it is valid, is that we are capable of performing only those actions whose necessary conditions obtain. Thus, when faced with the putative choice of performing action A which ensures an outcome consisting in some state of affairs X, or of performing action B which ensures an outcome consisting in some state of affairs Y, since X and Y are essential for their respective actions, only if X obtains can I do A, and only if Y obtains can I do B. Of course, if neither X nor Y obtain, I can do neither A nor B. If it is true that X obtains, it is false that I have a real choice as to whether I shall do A or do B. Since an essential condition for performing B is missing (namely Y) it never was on the cards that I could have done B. And similarly for action A if it is the case that Y obtains. If neither X nor Y obtains, obviously I shall not perform any action which ensures either X or Y. It is not up to me whether I perform A or perform B, because I have no power in myself to supply all those conditions essential for the occurrence of A and B — and I cannot *act* to somehow procure those conditions, since the conditions themselves are essential to my so acting.

I am not aware that there are any other ways in which the fatalist's doctrine can be argued for, and if there are indeed none, refutations of these two formulations offered by Taylor would constitute a refutation of fatalism. If I have made a mistake in this, then hopefully, any formulations of which I am presently unaware model either of the two above formulations sufficiently closely for my damaging remarks to apply to them as well.

Notes

1. Irving Thalberg, 'Fatalism Toward Past and Future'.
2. Gilbert Rye, 'It Was To Be', p. 15.
3. Richard Taylor, *Metaphysics*, p. 60.
4. Taylor, *Metaphysics*, p. 60.
5. See *Metaphysics*, pp. 62–4.
6. *Metaphysics*, pp. 66–7.
7. In 'Fatalism'.
8. 'Fatalism', p. 226.
9. Ibid., p. 223.
10. Ibid.
11. Ibid.
12. Ibid.

21
An Objection

At this point I ought to deal with the charge that in arguing against the fatalist, and adopting a libertarian position, I am maintaining inconsistent views, since if libertarianism is true, it is false that time is static — to be a *consistent* static theorist of time, the charge runs, I must side with the fatalist. As we have seen, in saying that what will be will be, it doesn't make any difference whether the events that *will* be really are tensed, or whether they are simply later than other events (and earlier than others): if fatalism is true, then the static theory of time can be true as well (although fatalism doesn't entail the static view of time, neither does it entail a tensed view — it's just that fatalism plus tensed time is consistent, and so is fatalism plus tenseless time). But what is not consistent, it might be objected, is libertarianism plus tenseless time, the very combination of views which I want to embrace. Can I logically get my way on this?

The reason for saying that I cannot, might run something like this. On the face of it a libertarian appears to be saying that there is a genuine metaphysical difference between past events and future events. What is true of all past events, that our deliberations and actions cannot now make any difference to what has happened and in this sense all past events are unavoidable, is not true of all future events, since there are some future events which are avoidable, because what we do now can make a difference to what happens later. Our present actions determine, to some extent, what is going to happen (in the future), but our present actions cannot determine, to any extent whatever, what has happened (in the past). There appears to be a real difference between the past and the future, given that we think the truth lies with the libertarian. The static view of time argues that there cannot be such a difference because really there is no past, present and future. It cannot be true, the static theorist seems compelled to hold, that an event at one time can be susceptible to our actions, but at another time not. The successful argument already deployed against the tensed view that events can be at one time future, at another time present, and at another time past, can be used with equal effect against the notion that events can be susceptible to human actions, or not, depending upon whether the events in question really are past or future. Since events never are either past or future, the conclusion has to be that there never are any times at which some events are susceptible to human actions and some times at which they are not, because their being susceptible is dependent upon their being future, something which is never true of any event. The libertarian seems to require that events can change with respect to their susceptibility to human action, but their changing in this way is logically dependent on their changing their A-determinations, and this latter change is ruled out by the static theorist's arguments. If events do not really change their A-determinations, they cannot change with respect to whether or not their happening can be influenced by human agents.

This charge can be met by maintaining a special formulation of the libertarian position, under which it is held that the future *is* fixed, yet nevertheless it is also true that, against what the fatalist urges, some things that happen to us and some things we do are avoidable, and that whatever we do (as distinct from what simply happens to us) we could have done, and can do, otherwise. It is, I believe, often up to me what I do; I am not fated to do just what I do. What I do is often the result of my own deliberation and my own subsequent free action. Against the fatalist, lots of things that I might do, and lots of things that might simply happen to me[1] are avoidable — all depending upon my skill in assessing differing situations, and in assessing and deploying my diverse abilities. What is true of me in this respect is true of everyone. My task now is to show that this is the case despite my belief that time is not tensed and that the future is not 'open'. If I can do this convincingly and comprehensively, without begging the question, never making appeal to tensed facts, but only to tenseless facts, I shall have succeeded.

Note

1. I am thinking here of examples such as my being blown by the wind off a cliff. We all know the difference between doing something and simply having something happen to oneself.

22
The 'Open' Future

Like anyone, when I turn my attention to the future and think of all the different things that might happen, I am aware of a mass of possibilities. Some things seem much more likely to happen than others, but the likeliness I place upon different events occurring provides only the vaguest guide to how the future will turn out. This is not something unique to me. One would be very foolish indeed if one took even a very well supported consensus that such and such will occur to *entail* that it definitely will occur. Even experts are frequently wrong. To live and act at all, we have to guess at what will happen, and as often as not our guesses are to some degree reliable. We each proceed by a kind of intuitive induction, some of us exercising this faculty better than others. We see as it were a picture of the future, composed of a blend of elements: expectations, fears, hopes. When faced with a decision which we believe will make a big difference to how the future will be, we might entertain several pictures of the future, each illustrating the expected outcome which we think will result from the different actions we can take. The pictures we have of the future are provisional. We accept that their accuracy is only partial, and that as things happen we have to update them continuously; yet having updated my picture of the future, it turns out to be no less provisional than it was before; some bits of it seem more certain than other bits, and my faith in this certainty is sometimes valid, nevertheless, the larger part of my picture of the future is vague and hazy. If I am honest about this, I shall have to say that in looking at my picture of the future I am looking at my own limited imagination. That 'realm of possibilities' which I peer into every time I wonder about and try my best to plan for what is going to happen is a creation of my own mind.

It seems to me that here we are faced with a question very similar to one we have dealt with earlier. We wanted to know what the metaphysical truth was behind our talk about the flow of time; and as it turned out, our talk is very misleading and does not reflect a genuine flow at all. What then is the significance behind our talk of the future as a 'realm of possibilities', as being 'vague' or 'indeterminate'?

I believe that when we talk about a future event in terms of possibility, as for instance we would be doing if we referred to the possibility of the Prime Minister's assassination at a public rally next month, we are referring not to a genuine metaphysical possibility that somehow or other at this moment of time resides alongside all those other possibilities which may or may not happen, but to the state of our own knowledge. We do not as a matter of fact know whether or not at the date in question the event of the Prime Minister's assassination takes place. With respect to the future, that we do not know what the facts are, about which events occur at which dates, does not validate the conclusion that the events are essentially merely 'possible', 'not really there', 'still indeterminate', or what have you. Our lack of knowledge is of course consistent with the alleged fact that the metaphysical reality of this matter is that the future is in some sense merely 'possible' or 'indeterminate', but our lack of knowledge does not entail that conclusion. After all, we lack knowledge about large parts of the past, yet no one suggests that sound logic is served by our concluding from this fact that these portions of the past are as yet indeterminate, that the events we may assume to have taken place at these times are still merely 'possible'. The conclusion that the future is somehow 'open' must rest on more than the fact that we are ignorant about most of the future, if not all of it. I do not believe that it is intuitively obvious that the future is 'open', as I shall point out in a moment. Those who hold that the future is 'open', I think, fear for their freedom. They think (wrongly in my view) that a closed future entails fatalism, when it is not more than consistent with fatalism.

I believe that all demonstrative statements, so long as they genuinely assert something that could be logically true of their subject, are either true, or false.[1] Thus, if it is said that it is possible that X is the father of the infant, we firstly mean that we do not as a matter of fact know whether X *is* the father. We also imply that he *could* be in that what we know of the situation (regarding perhaps details of his liaison with the mother, his relevant medical status, his attitude to contraception, and other details) makes it sensible to think that X is a good candidate for being the father. But whether he is the father is a determinate fact. When we say that *possibly* he is, whatever else we may be pointing to, we are pointing to our ignorance of a determinate fact. Again, it might be said that it is possible that the train crashed because the signal failed. Either the signal failed, or it did not, and we might know that indeed it did. And if it did, either it did or it did not causally contribute to the crash. 'Possibly' serves, amongst other things maybe, to mark our ignorance of the causal conditions responsible for the crash. I think that 'possibly' serves the same role in statements about the future. If someone says, 'Possibly the train will crash at 2.00 p.m. tomorrow', I believe that at that date there is either the event of the train crashing, or there is some other event which can be described as the train not crashing. 'Possibly' refers to the state of human knowledge, and not to a metaphysical reality.

Let us start with the set of events which could logically possibly happen in this world. A sub-set of events will comprise those that never have happened and those that never will. This sub-set we discard. What we are left with is the entire history of the world, all jumbled up as it were. These events can now be assigned to their proper dates, and what we finish with is the B-series. There are now no possibilities in sight. Everything is determinate. If we choose any time whatever, *t*, it matters not a bit whether that time is past to our now thinking about this, future to our thinking about this, or simultaneous with our thinking. Whatever event E is, it is either occurring at *t* or it is not. Let us suppose that it is occurring at *t*. If our chosen time, *t*, is future, someone might object, holding that E only *possibly* occurs at *t*, hanging on to the idea that the future is 'open'. So we wait until *t* and see what happens. We will take it that, when *t* comes round, E does in fact happen. But this view would entail that at some times E is determinately at *t* (namely at *t* and all times subsequently), but at other times (before *t*) it is not determinately at *t* (since 'possibly at *t*' implies 'not definitely at *t*' which entails 'not determinately at *t*'). This view I take to be incoherent, because I cannot see what it means to hold that there are times when events are at the times when they occur, but there are also times when they are not (since they are taken to be only 'possibly' at those times). It seems to me that we cannot ask, knowing that E occurs at *t*, 'Yes, but at what times is it the case that E is at *t*?' Events occur at particular times; it is not the case that events occur at particular times, *at particular times*. Other than saying this, I can see no further way to argue my claim. Though it can be added that an opponent who wishes to hold that events *can* possibly be at their dates at some dates, and actually or 'really' be at their dates at other dates can only mark the event's transition from the one status to the other by marking either its change, or its date's change, in A-determination from future to present. That analysis of what it means for an event to be at first possible, and later actual, I reject, because neither dates nor events can change with respect to their A-determinations: there is no 'now' at which moment events have the mantle of 'actuality' thrown over them.

Holding to the idea that 'possibly' marks human ignorance of determinate matters, a model of the future (indeed of the whole history of the world) which presents itself is that of the map containing uncharted regions. If someone points to such a map, indicating a location within one of the uncharted regions, and says that possibly there is a mountain there of such and such dimensions, either it is the case that there is such a mountain at that location, or it is not. 'Possibly' in this context marks a shortcoming in human knowledge of matters that in themselves are strictly determinate. The analogy between a map with uncharted regions and the history of the world breaks down when we note that it is always possible to find out (somehow or other) which features lie at which co-ordinates. But there is no analogous exploration that can be undertaken to discover which events lie in the 'uncharted' regions of the future. The fact that we are largely ignorant of what the future contains does not warrant the metaphysical conclusion that some events (namely future ones) are indeterminate and 'merely possible'. The urge to believe that the future is indeterminate stems from the need to make our metaphysical beliefs consistent with our strongest and most precious intuitive convictions. One of these is that human actions (or at least some actions) are performed freely by agents, that what happens to us is at least sometimes the result of what we decide and do, and that there is a point in trying to avoid unpleasant or dangerous situations. To act freely in this way, it must be the case that whatever we do actually do, we could have done otherwise. The fatalist says that we never can do otherwise, because we can do nothing other than what we are fated to do. Our intuitions about our freedom are thus opposed to the fatalist's doctrine — what the fatalist says must therefore be false.

The fatalist says that the future, like the past, is fixed, and therefore *that* must be false if we really are free according to our intuitions.

Note

1. The proviso is required to rule out statements that simply do not apply. It cannot be asked whether it is true or false that my older brother is taller than me, because I do not have an older brother; since there is no such person the assertion that he is taller than me says nothing of me which could be true or false — the common-sense expression 'it doesn't make any sense' is the correct response to such assertions. That which doesn't make sense is not a candidate for truth.

23
Truth-fatalism Refuted

These connections are unsound. I believe that we are free in the way our intuitions urge, but this fact does not entail that the future is really 'open'. I will show now the flaws in formulation (1) of the fatalist position which I outlined earlier, indicating that its being (sometimes) up to us what happens does *not* require an 'open' future and does not preclude time being tenseless.

There is one minor point which can be cleared away right at the outset. The fatalist points to what can be called a parity between the past and the future — the past and the future are equally fixed and determinate. Our views of human action should match this overall parity; that is, just as our actions are powerless to influence events that have already happened, our actions are equally powerless to influence what is to come. Our charge against the fatalist is that, in arguing like this, he is begging the question. If there is no overall parity between past and future, there can be no requirement that there exist a corresponding parity of human-powerlessness-towards-the-past with human-powerlessness-towards-the-future; claiming that there *is* a parity of human powerlessness is just another way of asserting the fatalist's conclusion that acting is never to any avail. What I am calling the overall parity between the past and future is put in jeopardy by asking whether we really want to deny that anything which happens now can influence what will happen in the near, or more distant future.[1] We presumably do want to deny that anything which happens now can influence what has already happened, in which case strict parity between past and future is maintained only if we hold the corresponding denial that nothing happening now can influence the happening of anything in the future — this amounts to a disavowal of all causation.[2] Taylor does not want to take this line, evidenced by his remarking that the sort of fatalism he adheres to is *not* that which says 'certain events are going to happen *no matter what* . . . regardless of causes'.[3] If causation has application in Taylor's fatalistic world, so long as he denies backwards causation, he must reject a strict parity between past and future. In which case there is no underlying, or deeper, parity with which human powerlessness towards past and future can be matched. The claim that we *are* powerless with respect to the future is just another way of rendering the fatalist's conclusion.

This is a minor point against the fatalist, because he can simply fall back upon his central concept of the body of truth, saying, 'Since it has always been true that Osmo would take Northwest flight 569 from O'Hare — this fact about what Osmo would do being contained in the body of truth about Osmo's life — this means that he could not have done otherwise. Considerations about causation don't enter into this at all. Regardless of what we think about causation, perhaps even that there is no such thing, whether or not we think that event A occurring at t_1 causes event B occurring at t_2, if the body of truth about the world contains the truths that A occurs at t_1, and B occurs at t_2, then those events happen at those times, unavoidably through logical necessity.'

This brings us to a discussion of the idea of the 'body of truth' in accordance with which, the fatalist argues, our actions (and indeed everything and anything at all) have to comply.

There is a problem with Taylor's Osmo example in that the body of truth appears to be responsible for its own content. This of course is the element that gives the story its ironic power and convinces us that Osmo really is helpless to escape from his fate. Taylor says God, the prophet, and the book are merely literary devices used for the purpose of illustrating his argument, and that the basic situation corresponds to the fatalist's view of how matters stand with the rest of us who have not discovered a book that contains the story of our lives. This seems unconvincing however: the fact that Osmo has discovered the book is surely not an insignificant matter that makes no intrinsic difference between his case and ours. For it is the book itself that has a profound influence upon his subsequent behaviour (in particular, his behaviour which contributes to the aeroplane crash in which he dies). In trying to prevent the prophecies in the book from being fulfilled, Osmo is provoked into acting in the very way that brings about the events described in the book. In this way the book, or the body of truth, actually plays a causal role in bringing about the events that make it the body of truth that it is (rather than some other). Now this is surely not what the fatalist claims is the case for us who do not discover such a book. If there is such a body of truth for all of us, it exerts no such influence over our lives as does the book in the example of Osmo. Yet if we were to remove this ironic, almost eerie, factor from the Osmo story, it would lose much of the point that it is trying to make.

I want to think about the problem of the body of truth by taking a particular truth from the whole set of truths we are calling the body of truth and seeing how it relates to that event which it describes. If we suppose for the purpose of argument that the details of Osmo's life pertain to the actual world, the body of truth contains the truth that 'Osmo boards Northwest flight 569 at *t*' because, at time *t*, Osmo in fact boards this flight and not any other, and those things which he could have done at *t* (if indeed he really could, which the fatalist says he couldn't) he does not do. One major flaw in the fatalist's argument arises from his misrepresenting the nature of the relation between Osmo's action and the truth which describes it. That this truth exists in the body of truth, says the fatalist, renders Osmo's action of boarding the flight inevitable and unavoidable. 'No man can avoid that which is truly described,' says the fatalist. The 'inevitability' pointed to here is stronger than that which may be seen in the following situation: Osmo decided to board the flight; his doing so satisfies, or will lead to the satisfaction of, some of his desires; he also believes all this is the case; that is, his desires and beliefs being what they are render his boarding the flight inevitable and unavoidable. No, this is not what the fatalist holds: his view is that its being true that Osmo boards flight 569 at *t* renders his doing so *logically inevitable, necessarily unavoidable*.

But the truth of this matter is that even though the body of truth does contain the truth that Osmo boards Northwest flight 569 at *t*, this being so does not logically compel him to get on the plane. I hope I can show that it does make sense to think that Osmo, at *t*, could have done something other than get on the plane.[4]

There are two spheres of discourse. One concerns such entities as arguments, statements, premisses, conclusions, and within this sphere we can say that things can be respectively, valid, sound, true, false, entailed, implied, logically sufficient or necessary, or logically inevitable. We may say for instance that the truth of this premiss, given that this argument is valid, entails or renders logically inevitable or unavoidable this conclusion which is thereby true also. The other sphere of discourse concerns events and actions, decisions, intentions, and beliefs. By far the most important concept employed in this sphere is that of causation. One event can cause another. Someone's desire against a background of certain beliefs is sufficient (without being logically sufficient) for a certain action. Events occur, or happen, rather than their being true. One event can imply another; for example, my turning on the light implies that the fuse was earlier mended; but this is not logical implication. When we say that these true premisses arranged in this valid argument form *make* this conclusion true, this is not the same sense of 'make' we appeal to when we say something such as 'His shouting at him like that *made* him drop his pencil.' Given these premisses in this argument form, the conclusion is made inevitable. But there is nothing inevitable about his dropping the pencil. Even though his being shouted at did make him drop the pencil, the pencil's dropping did not happen inevitably, since it was perfectly possible that he be shouted at, and that the pencil shouldn't fall; his being shouted at might have made him grip the pencil all the more firmly, for instance.

If we do not take notice of the fact that, even though we use the same expressions when operating in these two spheres of discourse, different concepts are being employed from one sphere to the other, language will lead us to incorrect philosophical conclusions. The fatalist can be accused of making this very mistake. One truth can necessitate another — the truth that x is red necessitates the truth that x is coloured, since the first cannot be true unless the second is also. But events do not necessitate other events — or rather, they neither do, nor do not, necessitate other events, because this concept of necessitation (which I am equating with logical necessitation and logical inevitability) cannot be used in the sphere of discourse about events. The light cannot come on unless the fuse has been mended, but the light's coming on does not *necessitate* the mending of the fuse. It is physically impossible that the light should come on without the fuse being mended, but it is not logically impossible. However, applying the notion of 'logically possible' to actual physical situations takes us nowhere at all. We wouldn't think very much of someone's grasp of these matters if, finding that the light won't come on they say, 'Well that's funny. It's *logically possible* that it should come on.' We would say that the fuse's being mended is sufficient (against a background of prevailing circumstances which are each necessary) for its being possible that the light come on. But that the fuse is mended is not *logically* sufficient for its being possible that the light come on. A mended fuse cannot *entail* anything. The truth 'The fuse is mended', in conjunction with other truths, may well entail the conclusion 'Therefore the light will come on', but the mended fuse itself cannot have a logical relation with anything (either other events or states, or truths), though it might have a causal relation with other events or states.

From the fact that truths sometimes necessitate other truths, it cannot be soundly concluded that truths can sometimes necessitate events in the physical world. But this is the connection which the fatalist calls upon to secure his argument: he says that the body of truth contains the particular truth that Osmo boards Northwest flight 569 at *t necessitates* Osmo's so boarding the flight. Truths do not necessitate physical events; if anything necessitates Osmo boarding the aeroplane, it cannot be the truth which states that he does board the aeroplane. The only conclusion that can be validly derived from the observation that the body of truth contains the particular truth that Osmo boards Northwest flight 569 at *t* is that Osmo, at *t*, merely *does* (not necessarily does) board flight 569. Osmo could have done otherwise in the sense that if he *had*, if he had gone for lunch in the airport restaurant for instance, the body of truth would record the fact that *that*, at *t*, is what he does. Osmo is free to take any of those courses of action which we may suppose any man at an airport is able to take. Whichever he does take, the body of truth will record that option. The body of truth cannot *make*, in any sense at all, Osmo board the aeroplane. The idea that the body of truth can *influence* events in time is as absurd as the idea that an accurate map of a certain territory somehow determines the landscape it describes.

Any appeal that the fatalist doctrine might have for anyone arises only if the mistake is made of thinking that truths can logically necessitate events and actions as well as other truths, and that events can be determined not only by other events, but by truths which describe them.

The other mistake which seems required if one is to feel tempted by the fatalist, is that of thinking that our actions are fated *at* particular times. Because Osmo was acquainted with the body of truth about his own life, he came to believe that whatever he did had *already* been fated to occur. If we think of ourselves as Osmo, we have to imagine that whatever we do now we shall later read about in *The Life of Osmo* — indicating that whatever we do has already been written about. Osmo failed to realise that far from the book determining his actions, it was his actions which rendered *The Life of Osmo that* book, instead of any other book which would have contained accounts of different actions and happenings. If Osmo had done something which was not recounted in the book, that would have shown that the book was not a faithful representation of the body of truth about his life; it would not have shown that he had miraculously avoided his fate, because there never was any fate to avoid. The truths in *The Life of Osmo* are consequences of events and actions (some being Osmo's), regardless of the fact that these events and actions occur at determinate dates whilst the body of truth just seems to 'be there' in some sort of timeless sense. That some bits of this body of truth might be known by some individuals (either before the events described happen or afterwards) does not interfere with the relationship that obtains between the truths and those events and actions which they describe. And this relationship is captured by saying that the body of truth (about Osmo's life, or the whole history of the world) contains the truths it does, because of what actually happens at each and every date which a complete body of truth will describe.

When the fatalist says that it was true beforehand that Osmo would board Northwest flight 569 at *t*, he is misrepresenting the relationship between truths and what they describe. It just is true (and is not true only at particular times, or even at all times) that at *t*, Osmo boards Northwest flight 569. It is true, because at *t*, that is what Osmo does. The point of saying this is to urge the view that truths cannot be assigned to dates; truths do not occur or endure, therefore they cannot occur at certain times or endure between certain dates. Truths are 'timeless'. That this is so entails that whenever someone utters a particular locution which as it happens expresses a truth about something, they would be saying something true. It does not follow that the truth itself is 'true at all times'. I do feel that people have been confused about this, and the fatalist plays upon this confusion to make this position seem correct. Now what definitely is the case is that people's saying things must occur at determinate dates. If I say something, that must be happening at some particular time. But if I say something which is true, it does not follow that the truth I have expressed is true at that time. If I assert the truth that '2 + 2 = 4' it seems to me ridiculous to hold that that is true on 3rd June 1987. 'Well if it's true at all (and it is) it is true at all times.' I take exception to the idea of 'being true' at dates, even all dates at once as it were.

Dates are when events, actions, realisations, dreams, statings, and what have you, occur, and when enduring states begin and finish. Statements which express truths are true at the time they are stated because this is the time the uttering occurs; it is not the time at which a truth is true, because truths are not in that category of entity which occurs or endures.

When someone states something that is true, the truth that is thereby expressed is neither true *at* the time of the statement, nor true 'at all times'. It is just true.

We could again speak of different spheres of discourse. There are those things which are dated, or have dates, which occur at particular dates or endure between particular dates; and there are those things which have no dates, which do not occur or endure at all. And that which has no dates should not be spoken of as occurring at *all* dates. If we start using the language and concepts of one sphere as if they automatically apply to the other, confusion will result.

That truths are timeless and don't have dates in any sense at all is easier to see with some truths than with others. The fact is that I am taller than Napoleon. But it clearly makes no sense to ask when this is true. One sort of truth we have spoken quite a lot about already: those expressed by B-statements — that I sneezed three days before I coughed cannot be dated. The sneeze has a date, and the cough has a date, that the first occurred three days before the second is certainly a truth about this world, but this truth is timeless; it has no date itself — *it* does not occur, although statements which express it might, and any of *those* which do occur would obviously be dated. Truths should obviously not be confused with expressions of truths.

All this is intended to encourage acceptance of the remark which refutes the fatalist's doctrine as expressed in view (1): truths are not the sort of thing that can determine or necessitate events or human actions, even if they are somehow known before the events or actions they describe occur. That anyone should be 'fated' to do something by its being true before he acts that he would so act could only be the case if truths are true at particular times. The belief that they are, confuses truths with entities which have dates, which occur at particular times. It is therefore false that, even if a certain truth is known and expressed at a time prior to the action it describes being performed, that the agent concerned was fated or compelled to act as he did.

The body of truth which the fatalist appeals to can in no way influence our decisions and actions. It seems reasonable to hold that the body of truth is somehow or other *made* to be this body of truth rather than any other. What makes it true that Osmo boards Northwest flight 569 at *t* is his so boarding that flight. Whatever I do at time *t* makes it true that at time *t* that is what I do. Other than saying things like this, I cannot see how else to elucidate the idea that the body of truth is *made* the way it is by what it describes.

These remarks, I hope, reveal the mistakes which the truth-fatalist makes. If anyone is to make us fear for our freedom, it is not the truth-fatalist. Before I say why I think fatalism as formulated in view (2) is also false, I think it is important to say more about my view that what agents do is up to them, and that (all things being equal) agents can do other than they do do, despite the fact that the future is determinate and closed.

Notes

1. Cf. Irving Thalberg, 'Fatalism Toward Past and Future', p. 29.
2. I am not prepared to accept that the *only* type of causation might be that between strictly simultaneous events, because if 'cause' is denied an explanatory role in our understanding which events result from which other events which occur before and after each other, we have no concept worth discussing.
3. *Metaphysics*, p. 59.
4. My remarks are not going to convince determinists and others who have independent reasons for rejecting free will; I am having to content my self that in this case, my one stone is going to kill just the one bird. The point being that if we are not free to do other than we in fact do, the fatalist has not got an account of why this is the case.

24
Acting on the Fixed Future

Even though the fatalist has no argument for its being the case that everything which happens happens unavoidably, it does not follow that agents can freely decide how to act and that their actions really do make a difference to how things would otherwise turn out.

My belief is that the view that human action is efficacious can be shown to be consistent with the view that time is static, despite the requirement of the static view that the future be fixed.[1]

Firstly, that there is a portion of the body of truth about the world which describes, what are for us, future events does not undermine the fact that an agent, who is about to act, can do other than that which he will do, which act is described by a truth in the body of truth. The correct view of this, as already anticipated in the previous chapter, is that whatever the agent decides to do and thus does will be recorded in the body of truth. The truth about what the agent in fact does is not a thing which occurs at a particular date or even at all dates, so that it is true (timelessly) that at *t* the agent does X says nothing at all about whether X was the result of a previous free decision to perform X as opposed to any available alternative. That the agent does do X does not entail that he couldn't have done something else; if he had, the body of truth would record that fact.

I think the reason why some people might see the fixed future as an obstacle to free action results from their insistence that to act freely agents must be able to *alter* what will be. But if the future really is fixed, altering any of it is ruled out. It might also be thought that if the future is fixed it makes no sense to suggest that in acting we are genuinely *choosing* between alternatives to make happen, or try to make happen, that alternative which we prefer.

Genuine choice is essential to free action, it would be claimed, but that the body of truth says that I shall do X at time *t* eliminates all possible alternatives to X which in order to act freely I must be able to choose between. And lastly, it seems to follow from the notion that the future is fixed, that all events in time are 'there', that all events 'exist'. If this sense of ontology is not extended to all events in history, we shall end up wondering *what* is fixed. If permanent and timeless B-relations hold between all events, it would seem that all these events must somehow 'be there'. If Osmo's boarding Northwest flight 569 at *t* is already 'there', it cannot make sense to hold as well that Osmo can genuinely choose to do something other than get on an aeroplane. It will be maintained that free action will be ruled out if the future is somehow 'there' already, if agents are not able to alter what will be, and if agents cannot choose between equally real possibilities.

My task now is to show that these considerations do not really speak against the compatibility of the fixed future with free action.

It is important not to confuse the idea of altering what will be with the idea of influencing what will be. What will be will be,[2] and that leaves us needing to decide whether it is true that human decision and action can make any difference to what will be. If it does make a difference, then some of our deliberations and actions can be said to influence future events. But influencing is not the same as altering. The claim that agents must be able to alter what will be is incoherent for a number of related reasons. The concept of 'altering' has its primary use in discourse concerning physical objects and people. Thus someone may be said to alter a physical object when they do something to it, such that at a time immediately prior to its being altered the object displayed a certain set of features, and that after the act of altering has been performed the object displays a set of features such that at least one feature is not what it was. That this feature is different must be put down to what the agent did to the object. (We can similarly alter a person's body, as well as their state of mind.)

If something is altered, we change it from how it was and how it would otherwise have been to something different, and this can only be done if the object or person being altered is present. That which is to be altered must already exist if we are to do anything to it to make it different. It therefore makes no sense for an opponent who objects to my view of future events as being fixed to maintain that we must be able to alter at least some of them. His alternative to the future being fixed is to say that it is not yet actual, 'a realm of possibilities', or something like that, and to describe the future in these terms is to make it something which necessarily cannot be altered, because it has no existence and therefore it has no features which human action could change. That is, someone who sees our freedom lying in the fact that the future is 'open' is denying that there is anything there *to* alter.

Not many people are likely to make this mistake about altering future events. By far the most important concept that needs investigating is that of choosing: it seems reasonable to hold that unless an agent can *choose* between various alternative possible actions, he cannot act *freely*. That someone might find themselves coerced by either a situation or a person threatening them into acting only one way such that they definitely could not have done anything else does not affect the discussion. What we need to know is whether freedom is impossible given the metaphysical facts of things. The metaphysical view that time is static and that the future is fixed will be resisted as long as it seems that this view rules out genuine choice for agents, which I accept is a necessary component for free action.

That I can choose which way to act from a range of possible alternatives does not require that those alternatives have any metaphysical reality. I wish to maintain that the possibilities I choose from have no other being than in my own imagination. They amount to thought, and that there should occur such thoughts in no way seems precluded by the future's being fixed. The question to ask is that of whether an event which occurs determinately at a certain time might so occur because of what someone did prior to that time. As agents, we all know that we can sometimes act so as to cause events which otherwise would not have happened at that time or in that way. The undeniable fact of our experience is that sometimes decisions we make result in bodily movements which themselves result in various effects in the world. No typing takes place unless *I* press the keys to make it happen. That it happens is entirely up to me.

'But this doesn't touch the problem, as I see it,' it might be objected. 'Let us imagine that a choice of three alternative actions is before me. Action A will cause event X, action B will cause event Y, and action C will cause event Z. Whichever action I perform will occur at time t_1, and the effect of that action will occur at time t_2. If we take the view that the future is fixed and determinate, that means that either X, Y, or Z occurs (timelessly) at t_2. This being so, I cannot see how it can make sense that now, at t_1,

I really can choose whether to do A, B, or C. If X occurs at t_2, then I must be doing A at t_1. There is no room for any choice.'

This objection rests on the mistake of seeing the effect of the action, this effect occurring at t_2, as *already* occurring at t_2, when the action which causes it is occurring at t_1. This error was discussed in the previous chapter. Let us imagine that the agent does A, which causes X. The reason why X occurs determinately at t_2 is because the agent did A at t_1. If the agent had decided to do B instead, it would be the case that Y occurs determinately at t_2.

'But how can any event occur at t_2 until what determines it has happened? Indeed, X may occur at t_2 because at t_1 I did A; A caused X. It seems to follow that X cannot be determinately at t_2 until t_1, when A happens.'

It is important to see that events are determinately at their dates, not determinately at their dates *at other dates*. It is wrong to think that X is at t_2 only at t_2 and ever afterwards, and it is wrong to think that, since action A at t_1 causes X, X is at t_2 *once A has happened*, but not before. Of course, A and X have dates; a certain B-relation obtains between them. That A is such and such earlier than X is timelessly true. The relation itself occurs at no time. Similarly, that A causes X is timelessly true. It is true that A occurs at a particular time, and X occurs at a particular time, but A does not cause X at a particular time. It is wrong to say that X cannot occur at t_2 because now (a time earlier than t_1 let us suppose) its cause hasn't happened. That would be similar to saying that X is not yet later than A: if we wait and see A performed, which results in X, we have to accept that one fact about the world which is true is 'A is earlier than X'. That we did not *know* this beforehand does not mean that the timeless truth 'A is earlier than X' was throughout the time of our ignorance not a truth, or that it became a truth only at t_1 when A happened, or at t_2 when X happened and A had happened.

It is up to us to make the future the way we want it. Future events, as well as past events, are determinately at their dates. The reason why certain events rather than others are at their dates (timelessly) is because of what some human beings, from our temporal perspective, have done, are doing, or will do.

Notes

1. And 'future' here ought to be cashed along the lines of 'all those

events which are later than any particular chosen event, including events which comprise changes in material objects, judgements, philosophers philosophising about time, any person's thought about later times, and other sorts'.

2. Saying this does not beg the question as to whether the future is fixed. What will be will be whether or not what indeed will be is now merely 'possible', or whether it is fixed.

25

The Difficulty of Finding a Model for the Static View of Time

If it is argued that all events are determinately at their dates, that, from where we stand, the past is fixed and the future is fixed, we must avoid thinking of events arrayed along the B-series like an object. Some people[1] have positively encouraged us to do this, talking in terms of the 'space-time manifold'. If we imagine time to be a fourth dimension of space, the movement and changes of three-dimensional objects in the world can be represented as four-dimensional objects which are completely static. On this way of thinking we are to imagine that a three-dimensional object as we ordinarily experience it is really just a cross-section of a much larger four-dimensional object. The whole world, with its entire history, can be conceived of as a very intricate four-dimensional object. An *event* on this model would be constituted by the intricacies in the four-dimensional object being a certain way at a certain location within the whole object. The event of a man being beheaded, for instance, would show up as the four-dimensional object which is the victim branching in two directions, one of these branches being his head, the other being the headless body. An object which changes colour would be represented as a four-dimensional object which is a certain colour in some parts, but a different colour in another part.

I object to this way of thinking because it harbours metaphysical confusion. The world we inhabit is not a four-dimensional object, but a mass of moving three-dimensional objects, doing those things we are all acquainted with their doing. To suggest that we can 'think of' the whole history of the world as a static four-dimension object implies that there is a viewpoint from which it can indeed be seen as a four-dimensional object. (Some might suggest that this is how God sees the world.) But if we decide to talk in terms of seeing an object, in terms of seeing that the four-dimensional manifold of the whole history of the world is such and such a way, this implies that that seeing takes place at a particular time, after which something else could happen. It is all too easy to think of God looking at this bit of the manifold, and then looking at a different bit: this being the case it would follow that event E occurring at time *t* (constituted by a particular bit of the manifold being just the way it is and not a different way) is *endlessly* the case, when we have seen that the correct way to describe matters is to say that it is *timelessly* the case. There is a difference. If something is endlessly the case, it logically might not be — it might be a certain way up until a particular time, then different from that, and this is not possible if something is timelessly the case.

In thinking of a four-dimensional manifold, we conjure an object which appears to endure and have a history. If we think this way, then event E, which is a part of the manifold, would have to endure as well, but its being the case that E occurs at *t* is not something that *can* endure or have a history. And it is most certainly wrong to conceive of the manifold as an unchanging object which stays the same for eternity, because if saying this is to have any content we must suppose a background against which it is possible that it *could* change.

There is not a problem here for the static theory of time. The problem is in trying to think clearly about what is truly the case about events and their dates. When we say that all events are determinately at their dates we all too easily picture the events as perhaps beads strung on a thread. For we want to picture the events as being there now. 'At this moment,' we are inclined to say, 'all the events of history are occurring at their respective dates.' Not true. Events *are* indeed (tenselessly) occurring at their dates, but they are not doing that at any time or other, not even now. That an event occurs at its respective date is not another event which needs to be dated.

Because all events are B-related (stand in B-relations) to all other events, we are inclined to want to say *when* they are so related, and as we have seen, this further question is not legitimate. Events do not have to be 'out there', 'happening at their dates even as we speak'. We must not think of events like we think of objects. If several objects are spatially related to each other they necessarily all exist now. The temptation is to say the same of events, that if they are temporally related, they must somehow 'exist now' for this to be true of them. But here we are mixing spheres of discourse. Objects exist, and that they do means that they are spatially related to other objects. Events happen or occur at dates, and that they do means that they are B-related to other events.

Thus I object to Smart, who would acknowledge the dangers I point to when thinking in terms of space – time manifolds, but says that it is all right to think this way if we think of the manifold like we think of a geometrical object of the sort mathematicians study, which does not exist or endure or have a history and is not perceived from any temporal viewpoint. If this can be done at all, it is done by thinking of an abstract object which has only mathematical properties. But the world is not an object like this. The mathematician's object is not an object at all. He uses the term 'object' to denote something which, to be the sort of thing he is interested in, must be very unlike what an object is in ordinary experience. The depth grammar of the mathematician's 'object' is different from the depth grammar of 'object' when used of everyday, ordinary physical objects. Ordinary objects endure, have histories, and have other mathematical properties. The manifold is not an object of this sort, because it makes no sense to say that it endures. And the manifold is not a mathematical object, and the objects which comprise it are not mathematical objects (what a way to think of ourselves!). The difficulty with what Smart says comes to this: the world we inhabit is made up of objects whose nature is to change continuously their non-relational and relational properties, which Smart proposes we should represent by a single unchanging object. For the reasons already offered, this cannot be done without introducing mistaken notions about time.

There is, in my view, no model which pictures the static concept of time. The truth is that events occur (tenselessly) at their dates, and there is no way human beings can get a picture of events doing this without making it look as though the events are all occurring at their dates *at the time the picturing occurs*, which we know is wrong because an event occurring at its date is not itself a dateable event which occurs at some time or other (such as when someone tries to picture static time).

Notes

1. Donald C. Williams, for instance, in 'The Myth of Passage'.
2. 'Spatialising Time'.

26
The Error in Taylor's Condition-fatalism

Before moving on to our final topic for discussion, I want to say why I think Taylor's condition-fatalism fails.

In the sphere of actions and events and conditions, when Taylor says that something is sufficient for something else (that ingestion of cyanide is sufficient for death, for instance) or that this ensures that, he specifically rules out the possibility of this relation being logical. We may recall his definition: '. . . if one state of affairs *ensures* without logically entailing the occurrence of another, then the former cannot occur without the latter occurring. Ingestion of cyanide *ensures* death . . . though the two states of affairs are not logically related.'[1] In the example of the poisoning, there is no logical connection between ingesting cyanide and dying. Disallowing logical relations between happenings I support, because the concepts of logical entailment and implication, of logical sufficiency and necessity, do not apply in the sphere of happenings. (This was discussed in Chapter 23).

What sort of sufficiency is Taylor appealing to? His examples, to my mind, point to causal sufficiency. I see no way of interpreting the statements 'Ingestion of cyanide is sufficient for death' or 'Ingestion of cyanide ensures death' other than by saying what is pointed to here is the *causal* connection between taking poison and dying. These statements mean 'Ingestion of cyanide *causes* death.' And similarly with the reverse concept of necessity. How else can oxygen be essential, or necessary, for the continuation of human life, except in the sense that oxygen causally contributes (along with other things)[2] to the mechanism of biological metabolism?[3]

Taylor's other examples of necessary conditionship are causal as well. That learning Russian is necessary to Taylor's reading a page of Cyrillic print means that the learning of Russian contributes causally to his exercising the skill of reading Cyrillic print. Similarly, when we say that having been nominated is necessary for S winning a certain election, we can mean nothing else than that the nomination contributes causally to the eventual victory: having been nominated causes certain officials to include the candidate's name to appear on the list which gets sent to the printer's, and that the name is on the list causes the typesetter to arrange the type so that when the plates are used they cause the candidate's name to be printed on the voting slips, and being thus printed contributes causally to each elector seeing that this person is indeed a candidate, and so on. And lastly, when Taylor says that having been in water is essential to one's swimming five miles, what can this mean except that certain training is causally required for the exercise of the skill of swimming? These are the only examples Taylor gives, other than that of the naval commander ordering a battle tomorrow, and it is this example which he uses to support the load of the fatalist argument.

There is only one way to interpret the statement that the commander's order for a battle tomorrow ensures a battle tomorrow, and that is that the order *causes* (all things being equal) a battle tomorrow.

Taylor's argument rests upon an equivocation between logical sufficiency (which was supposed to have been ruled out of the discussion), and Taylor's undefined sufficiency, which I am claiming can be seen as nothing other than the familiar concept of causal sufficiency.

The equivocation works like this. First of all, logical sufficiency and necessity: if A is sufficient for B, then B is necessary for A. For example, that X is red is sufficient for X being coloured; in which case X being coloured is necessary for X being red. The truth that A is sufficient for B entails the truth that B is necessary for A. This is the case when dealing with *logical* sufficiency. Taylor wrongly takes this rule to apply to his special undefined sort of sufficiency. He says that the naval commander's order, that there be a battle, ensures a battle, in which case, the occurrence of a battle is essential for the order that there be one. But having noted that we are really talking about the causal relationship, we can see that his rule is wrongly applied. That ingestion of cyanide is causally sufficient for death does not entail that death is causally necessary for ingestion of cyanide. The idea that a future circumstance can be causally necessary for something that happens now is difficult to understand. We understand that a *present* supply of oxygen is essential for the continuation of human life, but it's hard to attach any sense to the claim that a *future* supply of oxygen could be essential to our now continuing to live. As Thalberg points out,[4] we cannot imagine circumstances in which an impending shortage of oxygen can already be having lethal consequences. Similarly, our training successfully in the present to swim is not ensured (causally) by the fact that we later swim five miles, even though our currently being trained is essential (causally) for our eventual long-distance swim.

It seems plain enough that we cannot attach sense to the idea that essential conditions for A's occurring can themselves occur later than A. If we discover that the alleged essential conditions for A obtain at t_2, whilst knowing that A has already happened at the earlier time of t_1, we have sufficient grounds to reject the claim that those conditions were essential for A — that the conditions did occur later means that if they really were essential conditions for any event, that event cannot be A.

The naval commander, about to issue his order, is not on Taylor's view constrained by logic (although he is on the truth-fatalist's view). Taylor has made it clear in his definition that he doesn't want to discuss events and actions and conditions which are logically related; and, indeed, as we saw in Chapter 23, it would be a mistake to suppose that there can be any *logical* relations between such events, actions, and conditions, since the sphere of logical relations is confined to the sphere of premisses, arguments, truths, and such entities. This leaves the naval commander constrained by causation. (Unless I have unwittingly omitted a third alternative, which I do not believe is the case.) No event can take place if essential or necessary (causal) conditions are lacking. The naval commander's ordering whatever he does order, or anybody doing anything, I take it are just instances of events taking place. But, as I have argued, if we accept that the naval commander's ordering that there be a battle does ensure a battle, we would be mistaken if we for this fact allowed Taylor to beguile us into believing that the future naval battle is a necessary condition for the prior issuing of the order that there be a battle. It is just false that the upshot of a certain action has to be (or even can be) a causal prerequisite for that same action.

Notes

1. Richard Taylor, 'Fatalism', p. 223.
2. and thus does not alone ensure human life.
3. cf. Irving Thalberg, 'Fatalism Toward Past and Future', pp. 31 ff.
4. 'Fatalism Toward Past and Future', p. 31.

27
Death and Dying

At all those times prior to our births[1] we had no existence, and the truth of the matter is that each of us will assuredly die such that at all those times after our deaths we shall likewise have no existence. Some people of course believe that at death they will pass on into the after-life. 'Death' for those who think this means only death of the physical body. The person, who is conceived of as distinct from his body, survives. In other words, dying and death for such people are thought of merely as incidents in an otherwise continuous life. Such ideas about survival I am not here interested in. That does not mean that I think such ideas are definitely wrong and not worthy of discussion. I am interested to see what we *should* believe if we choose not to query the view that death is the end, that death of the body invariably means death of the person too.

Any human life is bounded at both ends by endless periods of non-existence. There is a philosophical puzzle to examine here, since people do not go through life concerned about their *earlier* non-existence, yet many people most certainly do go through life concerned about their *eventual* non-existence, indeed, it is not unknown for the unhappy few to be obsessed by the fear of death. My task in these remaining four chapters is to examine the concept of death and query the rationality of fearing death. Why *should* non-existence following death concern us whereas non-existence prior to birth does not?

Fear of death is sometimes, but not always, identified as fear of dying. Obviously death, in the sense of being dead, is not the same as dying. If someone has a fear of dying, they may or may not also have a fear of being dead. Dying is a state of going-to-die. Even though we are all going to die, we are not all dying. Going-to-die then, is a state a person can be in which will result in death; it is a state with a single empirically certain prognosis — death. Death is not inevitably consequent upon dying in the sense of going-to-die, although it usually is. A person may be dying from an illness yet be rescued at the last moment by a new miracle cure. Similarly, a person can be in the state of going-to-die because they have a severed artery, but the timely intervention by a doctor may get them out of that state. We may note also that a person can die even though they were not previously dying, as would be the case were a person (who is not dying) to be knocked down and killed by a lorry. And a person may die, but not from what they are dying of, as would be the case if a person who is dying from cancer were blown up by a bomb. In such a case, the *cause* of death would not be what the deceased person was dying of.[2]

The fear of dying, if what is feared is an unpleasant experience, is perfectly legitimate. An unpleasant experience is one which the subject has, such that his awareness that this is indeed his experience, comes in the reject-mode. (The idea of modes of awareness was discussed at the end of Chapter 18.) The thought that we are going to experience something unpleasant, itself comes in the reject-mode. So if prior fear of having a painful experience at the dentist's (say) is intelligible, fear of dying certainly is as well, since dying is just one out of a number of ways that one may encounter misery. We can be miserable now about the fact that we will later be in misery. Dying, even if it is nothing more, is obviously a source of misery to those who suffer pain and unpleasant sensations when they are in the state of going-to-die.

Many people would say that I have missed out the most important thing about dying. They would say that even though dying can be unpleasant because of the sensations that can occur while one is dying, much as a visit to the dentist can be unpleasant if we have to have a painful filling, say, it is altogether wrong to model dying on unpleasant experiences one can have during life. The fear of dying, they would say, wouldn't be that terrible if dying really were like going to the dentist, or even like being tortured; what is so terrible about dying is that it leads to death. Dying would be quite endurable really, if one could go on living once it was over. But one stops living when dying is over, and that's why people fear it.

The extra fear which is being pointed to here, over and above the fear of having unpleasant sensations whilst one is dying, is simply the fear of being dead, of death itself. My concern is not with dying, with that state of going-to-die, but with death in the sense of being dead. 'Being dead' merely categorises someone's non-existence occurring later than their being alive. There are a number of considerations which urge the view that fearing death is not as rational as many people like to think it is; one can go quite a long way in defending this view, and the objections which arise in pursuing this line of argument to my mind do not establish conclusively that there is no hope for the irrationality of fearing death.

Notes

1. 'Birth' here should not be taken as the moment at which we come into the light of day. It seems true that before the moment at which that happened to me, I nevertheless existed. It seems that there is no sharp boundary dividing non-being and being for people. Even conception takes *some* amount of time. It seems arbitrary to fix a precise date for a person's coming into existence. Yet people do come into existence, and there are times at which we did not exist. Coming into existence, for people, is a 'fuzzy' business. A simlar observation can be made about ceasing to exist after life.
2. cf. Ninian Smart, 'Philosophical Concepts of Death'.

28
The Evils of Death

Whatever we may think about whether S's death is an evil for S, S's death obviously might be an evil for others. This aspect of death, which we are not interested in here, can be acknowledged and discarded.

Whatever S's feelings for the other people concerned, it is perfectly rational that others should regard the death of S an evil for themselves. This hardly needs spelling out. S's spouse, family, friends, and work-colleagues are all likely to view S's death as a misfortune for themselves. The nature of the misfortune is likely to be different depending upon the relation had with S. For instance, S's spouse is likely to experience S's death as more of a misfortune than S's work-colleagues though different cases will of course vary. And the sort of misfortune will be different from person to person. Upon S's death, S's friend may lack a golfing partner, and it is conceivable that being a golfing partner constituted the larger part of the relationship between these two people. The misfortune S's friend suffers has an altogether different character from that which a work-colleague suffers if S's death means for him merely that he will have to do, unwillingly, S's work as well as his own until a replacement for S can be found. (It should be noted that S's death can be an evil for me even if I have never met S, and S has no knowledge of me. This would be the case if for instance S is a novelist whose work I admire. It is reasonable for me to regard S's death as a misfortune for myself, since his being dead means that I can no longer look forward to the stimulation of his forthcoming work. Similarly, the death of a gifted politician could be regarded as a misfortune for everyone in the nation.)

Unselfish concern for the well-being of others will usually form part of the fear of death which someone may have. But presumably it would be said that fear of death is not constituted wholely by such concerns. For were it to be, fear of death would rank alongside fear of being physically handicapped in various ways, fear of being imprisoned, or fear of total amnesia, and all those fears of things which were they to happen to us would impose some measure of misfortune upon those who matter to us. But the fear of death adds up to more than unselfish concern for others who would suffer as a result of one's death. There is a selfish component as well. The fear of death is in part concern for oneself. Were someone somehow or other given the choice between death and amnesia, it seems reasonable that they might choose amnesia. If asked to account for their choice, they might say that those who matter to them would suffer less misfortune if they became amnesiac rather than if they died. But it would seem no less plausible if we were told that they selected amnesia over death for their own benefit. It seems intuitively sound to say that, for many people in most circumstances, death is a much worse fate than many other calamities including amnesia, disability, imprisonment, injury, and others. Being imprisoned is bad enough, for the prisoner, for those who matter to him, and for those who care for him, but death would usually be taken to be much worse.

It might be said that fear of death is to be cashed in terms of fear of our cherished projects being curtailed. If S is engaged on writing a long novel, say, his death would put an end to his realising his dream of getting it completed. The thought of this catastrophe constitutes his fear of death. This view is not right, if only for the fact that it implies immunity to fear of death can be had by those who have no projects which death would otherwise have cancelled, or by those who manage to complete current projects but who do not engage in further projects. The truth is that people without projects to complete fear death, if they do, no less than those with projects who also fear death. Apart from that we can make a similar observation to that of the previous paragraph, and say that most people would not rank death with other calamities which they may have befall them. One such calamity, for example, may be a certain sort of stroke. We may imagine S, his long novel still not completed, suffering a stroke which affects his faculties only to the extent that he can no longer write or speak English. If his fear of death could properly be couched in terms of his fear of his novel not getting finished, this disaster would be for him as bad as death itself. Indeed it may be: he may now quite rationally opt for suicide. But equally it may not. There is no difficulty in understanding that S should prefer his stroke to out-and-out death. We can note as well that many people would give up their projects if that meant they didn't have to die (yet). For many people, fear of death appears to consist in something over and above the concern that one's projects will be curtailed by one's death.

We will now look at the question as to whether S's death can be an evil *for S*.

29
Can Death be an Evil for its Subject?

Wittgenstein said that 'Death is not an event in life: we do not live to experience death' (*Tractatus* 6.4311). In other words, for anyone, either they are alive, or they are dead; if they are alive, they can have experiences, if they are dead, they obviously cannot have experiences. Experience of one's own death, in the sense of having an experience of being dead, is necessarily ruled out. Being dead would appear to be a misfortune which people logically cannot suffer.[1] Thus, being dead is a state which we can never ascribe to ourselves. It is logically impossible for someone to judge to themselves, 'This awareness of being dead comes in the reject-mode.' For each of us, our experience is that it is only other people who actually die.

The attempt to see death as an evil for its subject is thwarted immediately if we try to think of death in the way that we think of other evils. The sorts of evils that people can suffer are distressingly numerous. Someone may be physically injured or ill, they may be ridiculed by others, they may lose their wallet having just filled it up with money, they may lose their jobs, and so forth. Having such things happen to one involves having an unpleasant experience, feeling upset or 'nasty' inside. Being aware of these events as misfortunes means experiencing them in the reject-mode. Yet we reject some misfortunes more than others. For most people, having flu would be better than losing one's job, whilst both are nevertheless misfortunes: losing one's job *and* having flu is presumably worse than suffering just one without the other. That this is how people experience their lives justifies thinking in terms of a 'hedonic scale'. If we wanted to, we could each give a hedonic value to the experiences we have. Experiences we liked having a lot would go at the top of the scale, whilst the worst misfortunes would go at the bottom of the scale. Roughly, those experiences listed in the upper portion of the scale would be those we have in the accept-mode, and those listed in the lower portion would be those we have in the reject-mode. In the middle would be listed experiences for which we have neither a positive liking nor a positive aversion.

If death is an evil, it is plainly not an evil of the sort we all suffer from time to time, which we would place towards the lower end of the hedonic scale. Our being alive is essential to our experiencing *anything*; in order to put an experience on the hedonic scale we must be alive. That we can intelligibly make use of this hedonic scale, ranking our experiences according to their hedonic value, is another way of showing that people are capable of making value-judgements about what happens to them. 'This is good', 'This is bad', 'This is a misfortune', are statements which can be used to express the value we find in our experiences. If someone claims that death is an evil, I see no way of taking that claim other than that death can be evaluated, and that the evaluation death gets awarded is low on the hedonic scale. But since death is not something people are capable of experiencing, it seems hard to accept the intelligibility of 'Death is an evil (for him who dies)' as a value-judgement. The claim that 'Death is an evil' is a value-judgement requires that it has to be judged *by someone*. This claim is problematic because the subject of a death cannot judge it himself to be an evil — he is beyond judging anything. Things have value only from personal standpoints: in which case, where does S's death get its value of 'evil' *from*? Who, other than S (if we are interested only in S-relative values), can make the judgement? One solution to this puzzle is to say that 'Death is an evil (in S-relative terms)' is not really a value-judgement. That surely is the wrong route to take, since the aim of our discussion is to understand the rationality (or otherwise) of the fear of death; but how are we to understand fear of anything except in terms of the object of fear having for the fearful person a negative value? To deny that 'Death is an evil' embodies a value-judgement is quite obviously wrong. The claim that 'Death is an evil' is a value-judgement, if what is being pointed to is the misfortune of living people who suffer in consequence of someone's death is perfectly acceptable. S's friend can place S's death on his hedonic scale as well as he can place other things happening in his life. Our reaction to those who say that death is an evil for *him who dies* should be to ask how that value-judgement is arrived at. S's death is not something S can evaluate on his hedonic scale. That being so, what grounds have we got for saying that S's death is an evil *for S*?

The misgiving I am pointing to arises because death is not something which a dead person experiences. The dead person is therefore not in a position to get a hedonic rating for death. The value-judgement 'S's death is an evil for S' doesn't seem to be saying anything intelligible.

More can be said about this, which will help to elucidate these ideas. The following discussion, which is supposed to enlarge on the foregoing, is derived in main from Part I of Harry S. Silverstein's article, 'The Evil of Death'.

The claim whose intelligibility is being queried is:

(A) S's death is an evil for S.

This statement is 'S-relative' because it says that something is an evil *for S*. This is unproblematic, since if anything is to be an evil, it has to be so *for* someone. What does S see himself doing when he says that death is an evil for himself? He would not want (A) interpreted in one way which is possible, that when he says that his death is an evil he means that this particular death (that is, this particular way of dying) is worse than other alternatives (though on other occasions he might mean this). (A) is not meant to be taken here as a death-to-death comparative claim. It is because life is so much better than death that death is an evil; (A) is, in other words, a life-to-death comparative claim, and must be interpreted as asserting, in part:

(B) S's death is worse for S than S's continued life is bad for S.

(These two assertions do not mean the same, because (A) implies, obviously, that S's death is an evil for him, whereas (B) does not imply that.)[2]

Unless we do interpret (A) as making a comparison, either a death-to-death one, or a life-to-death one, it's hard to see what the assertion should be taken as saying. The death-to-death comparison is not the one that interests us here. But (B), the life-to-death comparison, demonstrates the unintelligibility of regarding S's death as an evil *for S*. For how is the truth of (B) to be established? Since our interest is in S-relative values, the only person entitled to assert that (B) is true is S himself. But S is never in a position either to know the truth of (B) or to assert that truth. (B) asserts a comparison which logically cannot be made. S cannot make the comparison when he is alive, because at no time can he have experiences which reveal the sort of value being dead should be awarded. The other side of the comparison he presumably can evaluate: if a continued portion of S's life is bad, he can place the experiences he has at this time on his hedonic scale. At best this enables him to make a life-to-life comparative claim, viz., 'This part of my life is worse (for me) than other parts of my life have been bad (for me)', or 'This part of my life is bad and is no better (for me) than any other part has been.' The view that death is an evil for him who suffers it remains unintelligible.

It might be said that there is a different starting-point which reveals matters more in accord with the death-is-an-evil view. Instead of saying that death is a positive evil, we should say that death is the lack of a positive good, being alive. Does it not follow from the fact that it is intelligible to regard S's life as a good that its loss can be intelligibly regarded as an evil? Saying this confuses the two ways of seeing S's life as a good and S's death as an evil. S's life may of course be a good for others, and his death may be an evil for others. But our interest is in S's life being a good, and S's death being an evil, *for S*. Saying that S's being alive is a good *for S* is no less problematic than saying that S's being dead is an evil *for S*. This is so because being alive cannot rank alongside other goods that S may receive and it cannot have a position on his hedonic scale. Being alive is a prerequisite for S's receiving either goods or evils. If being alive was simply a good like other goods we would be able to compare the value of being alive (for ourselves in self-relative terms that is) with goods such as eating ice-cream or reading a good book, and with evils such as having food-poisoning, or being robbed. But it is not intelligible to claim that being alive is either better or worse than eating an ice-cream, and to say that having food-poisoning is worse than being alive (which it would have to be were being alive really to be a good) is to utter nonsense. That it is true of S that he can be alive, and that it is true of S that he can have food-poisoning, it does not follow that S can weigh being alive against having food-poisoning. And we do not need to regard being alive as a good to understand that having food-poisoning is an evil.

The claim:

(C) S's life is a good for S,

has both a life-to-life comparative interpretation, and a life-to-death comparative interpretation. The life-to-life interpretation is that the life S as a matter of fact has is better (for S) than, either how it was before, or how a different possible life would be (for S). This interpretation we are not here interested in. Interpreting (C) as comparing life to death means that it asserts, in part:

(D) S's continued life is better for S than S's death is good for S.

This assertion is similar to (B), in that it too asserts a comparison which logically cannot be made. Its truth depends upon the S-relative values assigned to S's continued life and to S's death. As before, S's death is something which S himself cannot evaluate since S's not being dead (his being alive) is necessary to his evaluating anything at all.

Seeing the issue of the fear of death in this way urges the view that fear of our own deaths is irrational. Our non-existence once our bodies expire should concern us no more than does our non-existence prior to our bodies being constituted. The attitude of fear and concern which many extend towards future non-existence is not justified.

Notes

1. cf. also Epicurus' observation in *Letter to Menoeceus*, quoted by Harry S. Silverstein, 'The Evil of Death', 'So death, the most terrifying of ills, is nothing to us, since so long as we exist, death is not with us; but when death comes, then we do not exist. It does not concern either the living or the dead, since for the former it is not, and the latter are no more.'

2. That X is worse than Y, for S, does not entail that X is an evil, for S. For instance, S's winning merely half a million pounds could properly be construed as worse for him than his winning a million pounds, but it does not follow that winning half a million pounds is an evil.

30

Objections to this Analysis

The analysis offered in the previous section rests on the assumption that if something, X, is to be an evil for S, S's experience of X must be unpleasant. For how can X have a value for S ('evil' let us say) if S does not have an appropriately unpleasant experience of X?[1]

That an S-relative evil must result in S's suffering, has been queried by Thomas Nagel, who points out that this assumption finds expression in the everyday remark 'What you don't know can't hurt you.' He offers an objection by way of a counter-example, to thinking this way. If it is true that what S doesn't know about can't harm him, it would follow that even if S 'is betrayed by his friends, ridiculed behind his back, and despised by people who treat him politely to his face, none of it can be counted as a misfortune for him so long as he does not suffer as a result'.[2] Of course S would be aware of his misfortune if he discovered the betrayal, and would presumably feel positively unpleasant in consequence of this discovery. But what the example is supposed to urge is the view that something can be an evil *for S*, even in the circumstances where S is unware that the something is going on and does not suffer a positively unpleasant experience as a result of its going on or his knowledge that it is going on. The view under attack, that X is an evil for S only if S has a nasty experience of X, would consider betrayal (for instance) to be bad because betrayed people feel bad when they learn of their betrayals. Nagel thinks that his view explains why the discovery of betrayal (and other misfortunes) causes suffering in a way which makes suffering on such an occasion reasonable:

For the natural view is that the discovery of betrayal makes us

> unhappy because it is bad to be betrayed — not that betrayal is bad because its discovery makes us unhappy.[3]

The claim that betrayal is bad needs to be examined. ('Betrayal' could be replaced in the discussion by any other evil, such as deception, robbery, libel, and any others which it seems one might suffer without having a positively bad experience at the time of their occurring.)

Nagel speaks in terms of explanation. For him, that betrayal is bad explains why the discovery of betrayal makes us unhappy. What I wish to dispute is the claim that betrayal is bad if this is interpreted the way Nagel interprets it, which is to say that betrayal is unconditionally bad (that is, not bad conditionally upon the subject's suffering). He seems to be saying that it must be the case that betrayal is bad (unconditionally) because we need this fact to explain why it is reasonable to feel distressed when one discovers that one has been betrayed. The discovery of betrayal makes us unhappy because it is bad to be betrayed. Betrayal being bad is a sufficient condition for feeling unhappy when betrayal is discovered. In other words, what Nagel maintains is the conditional statement 'If betrayal is bad, then one feels unhappy when betrayal is discovered.' There is no way that the conclusion 'Betrayal is bad' can be validly derived. It is reasonable to affirm the consequent: the truth is that people do feel bad when betrayal is discovered. But affirming the consequent does not help us establish the truth of the antecedent, that betrayal is bad unconditionally.

I take the fact that people feel bad upon the discovery of betrayal to be something that is given. Why does Nagel feel the need for a further explanatory fact which would make it 'reasonable' to feel bad about finding out that one is being betrayed? (Similarly, if we look at our aversion to being ridiculed, say, the given fact is that people feel bad if they know they are being or have been ridiculed. Saying that 'Ridicule is bad (unconditionally)' would be taken as shorthand for saying that people feel bad if they find out about, or know about being ridiculed; it seems hard to see that anything is explained by pointing to a further fact, that ridicule is bad.)

I see this the other way round from Nagel. What needs explaining is why we say 'betrayal is bad'. There are two ways of answering this. We say that betrayal is bad *because* generally one feels bad when one knows, or finds out, that one is being betrayed. The only reason we have for ever saying that X is bad is that people experience X to be bad. That people *feel* bad when they are betrayed supplies the reason for *describing* betrayal as bad. The second reason we can find as to why we say that betrayal is bad points to the fact that 'betrayal is bad' need not necessarily describe betrayal — it says also that there is a moral injunction against betraying people. Most uses of 'betrayal is bad' incorporate the descriptive element and the prescriptive element. The only reason we have for prescribing against betrayal is that people have bad experiences of it.

If someone is betrayed such that they neither know about it nor suffer any misfortunes as a result, it is nevertheless rational to tell the betrayer that his betraying actions are wrong because firstly it is immoral to act in a way which jeopardises someone's well-being (and his action does just that, since it is possible for the betrayed person to discover the betrayal and feel bad about it) and secondly, that this is the case is why there is a moral injunction against betrayal, and such injunctions should not be contravened.

The general point I wish to emphasise is this: people may generally have had bad experiences of X, which justifies asserting that 'X is bad', that is, X is bad because experiencing X feels nasty. If experiencing it didn't feel nasty, we would have no reason to say that X is bad.

Lastly, on this point, I object to the idea that people can have misfortunes yet also suffer no positive unpleasantness. If X is an evil *for me* that is so because I judge myself that this is the case, and I can do that only if I have an unpleasant experience of X.

I doubt that everyone would be convinced by my response to Nagel. It is going to be said that, true, part of betrayal being bad consists in the fact that when people know that they are being betrayed or when they find out that they have been, they feel bad about it and count it as a misfortune. This being the case is what justifies betrayal being an evil for a man even if he has no knowledge (yet, or ever) that he is the victim of betrayal. If he did know, or if he finds out about it, he would count it as an evil, and that means that it is an evil for him even if he stays ignorant. My answer to this runs: certainly, at the time he finds out, then the betrayal is an evil for him, but was it an evil before that moment? If something is an evil for a man, must it not be so at some particular time, or between some particular times, of his life? Betrayal, for a man at times before he finds out about it, is simply not affecting him, so can't be affecting him nastily[4] — we are back to wondering why, at this time, it should be counted an evil *for him*.

But if I may briefly summarise this digression into betrayal, I see matters this way. The view that an S-relative evil must result in S's suffering faces objections which to my mind are not overwhelmingly convincing. It would be a mistake to abandon the view without more damaging argument. Since S's death does not cause S suffering at those times when he is dead, we are lacking grounds for calling his death an S-relative evil.

What about cases where people seem capable of suffering an evil which results in *no* positively bad experiences, and where even the possibility of discovering that the evil has befallen them has to be ruled out? If there can be such evils, the principle that X can be a misfortune for S only if S has an unpleasant experience of X can be denied. Our interest is in whether death might be an evil of this sort. Nagel offers a specific instance. We are asked to consider the cases of an intelligent person who becomes the victim of brain damage, and is thereafter mentally comparable to a contented infant.[5] The brain damage is surely an evil for this person, yet they do not suffer as a result, and they are never able to find out about the evil that has befallen them because the very nature of the evil precludes this.

Certainly it is right to regard the brain damage as an evil from a number of points of view: if S is brain-damaged then his friends and relations have for all intents and purposes been bereaved of the person they knew. And it is likely that S's projects will suffer, since he now lacks certain abilities required to continue them, and harm could conceivably result (if, for instance, S was working on a cure for a certain illness, victims of the illness may now have to endure pain which otherwise was expected to be relieved). But does the evil extend to S himself? Is the brain damage *for him* an evil? If we look into what S's life might be like from S's subjective point of view it seems we have reasons for saying that the brain damage does not constitute an evil for S. What is the misfortune *for S* if before his injury he is often depressed and dissatisfied with his life because his ambitions are slow in coming to fruition, if he spends his energies overcoming difficulties, but after the injury he is happy and carefree, enjoying to the full the fruits of his new life such as they may be — eating, playing with children's building bricks and poster paints, splashing with water, and so forth?

The evil for S which Nagel sees is generated by the contrast between S's new condition, and what would have been the case if S hadn't been injured. S's post-brain-damage life is not far different, let us suppose, from that of a chimpanzee, and of course it cannot be regarded as an evil for the chimpanzee that it lives the life of a chimpanzee. But it can be regarded as an evil for S that he now lives the life of a chimpanzee because he isn't a chimpanzee, he didn't previously live such a life, he didn't choose to live such a life, and he wouldn't now be living such a life but for the brain damage. Living life as a chimpanzee means for S that he has suffered a serious deprivation, which is not true of a chimpanzee who started out as a chimpanzee.

Nagel's understanding is that even though S cannot make the life-to-life comparison

> (E) Being brain-damaged is worse for me than my continued living without being brain-damaged would be bad for me,

his suffering brain damage is nevertheless an evil for S. Against which we can answer that the delights of water-splashing and suchlike rank so highly on S's hedonic scale that were he somehow able to make the comparison, (E) would be judged false.

The difficulty for me is seeing a useful connection between these thoughts about brain damage and how we should regard our fear of death. Brain damage, as Nagel suggests, constitutes a deprivation or loss which means that various possibilities for S's life are cut off from him. Being imprisoned, for instance, creates a similar deprivation, only here (all things being equal) the prisoner is aware of his deprivation.

Even if we accept for the moment Nagel's view that brain damage is a deprivation, and is for that fact an evil, we are still no nearer to seeing death as an evil for him who dies, for these reasons. If indeed death is a deprivation, we could classify it with brain damage and imprisonment, and since we have here a set of eventualities which constitute misfortunes for those in receipt of them, we would know that death is a misfortune. But it is not at all clear that we should regard death as a deprivation. If someone is deprived of their liberty, we still have the person who has been deprived. Their being deprived in this way involves their having a certain sort of life prior to imprisonment, but a different sort subsequently. The same is true for brain damage. But death cannot be modelled on deprivations like imprisonment because the alleged deprivation of death does not derive from the victim's living a certain sort of life up to a certain point in time, and then living a different, deprived, sort of life thereafter. The truth that all deprivations are evils for those who suffer them is not very useful to us if we wish to see whether death is an evil for him who dies if we cannot establish that death is a deprivation.

To conclude this chapter, there are a number of other points to mention.

(1) It might be said that since experiences had in the accept-mode are good, is it not intelligible to regard what is necessary for the having of such experiences, that is, being alive, as a good also? The opposite of being alive, being dead, is therefore intelligibly regarded as an evil. This view unforunately cuts both ways, for we may just as well say that if experiences had in the reject-mode are evil, it is intelligible to regard what is necessary for the having of such experiences, that is, being alive, as an evil also. Thus, being dead, the opposite of being alive, is intelligibly regarded as a good. What is at fault is the principle the argument rests on: that if Y is a good, X is also a good if X is necessary for Y. One counter-example will suffice: S may regard going out to visit friends as a good, but it is reasonable for him to regard walking there through a downpour (this being necessary to his making the visit) as an evil.

(2) Someone, S, may try to justify their fear of death by claiming very simply, 'I just want to go on having experiences.' But this says merely that S wants to be alive, since being alive for human beings consists in having experiences.[6] Being alive is not being dead. S has stated merely that he does not want to be dead. He has not explained why it is reasonable to regard being dead as an evil.

(3) A similar point is made if someone says that what they fear in death is permanent unconsciousness. Temporary unconsciousness is not ordinarily feared; everyone is temporarily unconscious from time to time whilst they are sleeping. But what is so awful is the thought of being unconscious for ever. But this begs the question as to why death is to be feared. Death can be described as permanent unconsciousness. Someone who says that they fear permanent unconsciousness is saying no more than that they fear death. That this fear is justified is not discussed.

To summarise: my understanding is that events are good or bad depending upon the pleasure or pain they produce in those who experience them. Being dead results in neither pleasure nor pain for its subject, and it is not to be classified with those items which can be evaluated as good or bad on a hedonic scale. Though S's death can be a misfortune for others, it does not follow that it is a misfortune for S. If something is an evil, it must be an evil *for* someone, and we saw in the previous chapter that S's death being for *himself* an evil lacks intelligibility. Until I hear more convincing arguments to the contrary, I shall regard the asymmetric attitude of concern and fear which is directed at posthumous non-existence but not at prenatal non-existence to be irrational.

That one day, some time later than the present, an event will occur which is described as my death, after which time the person I know myself to be will never have any more existence, cannot alter its being the case that my life will have been led in such and such a way, that at these times I enjoyed these benefits and these joys, and at these other times I endured these miseries and these sorrows. Other than that my life has boundaries, a beginning and an end (which logically need not be the case), I can see no further significance in the fact that at quite a lot of times, possibly an infinite number, I have no experiences, some of these times occurring earlier than my birth, and the others occurring later than my death.

Notes

1. We should allow for X's being an evil for S if S has no direct experience of X, but does have a direct experience of Y which is unpleasant, and X causes Y. This seems to open the possibility of X's being an evil for S even though S does not know it, as would be the case if S is ignorant of the causal relation between X and Y. The scope of our discussion will not address this further complication. S's death obviously cannot be an evil in the sense pointed to here, of being the cause of something which S experiences in the reject-mode, since being dead, S can have no such experience.
2. Thomas Nagel, 'Death', pp. 4–5.
3. 'Death', p. 5.
4. We must allow for a man being adversely affected by the consequences of his being betrayed even though he is yet ignorant of the betrayal itself. He would then be aware that he is suffering an evil, but would not know its proper source. (I am thinking of say a salesman who finds his old customers no longer buy from him because someone has betrayed some terrible secret about him so that they don't trust him any more. The salesman suffers an evil, but does not know this results from his being betrayed.)
5. 'Death', pp. 5–6.
6. It might be said that one can be alive yet be in an irreversible coma. But it is obvious that such a state is as good as death itself, since one could not rationally choose irreversible coma followed by eventual death some time later over death right at the start.

Bibliography

Black, M. 'The "Direction" of Time', *Analysis* 19, no. 3 (1959) pp. 54–63

Bouwsma, O. K. 'The Mystery of Time (or, The Man Who Did Not Know What Time Is)', *Journal of Philosophy* 51, no. 12 (1954) pp. 341–63, reprinted in O. K. Bouwsma, *Philosophical Essays* (University of Nebraska Press, Lincoln, 1969), pp. 99–127

Braude, S. E. 'Tensed Sentences and Free Repeatability', *Philosophical Review* 82, no. 2 (1973) pp. 188–214

Broad, C. D. 'Ostensible Temporality', in Gale, *Philosophy of Time*, pp. 117–42, reprinted from C. D. Broad, *An Examination of McTaggart's Philosophy*, vol. 2, part I, Cambridge University Press, 1938

Christensen, F. 'McTaggart's Paradox and the Nature of Time', *Philosophical Quarterly* 24, no. 97 (1974) pp. 289–99

——— 'The Source of the River of Time', *Ratio* 18 (1976) pp. 131–44

Finch, H. Le Roy. *Wittgenstein — The Later Philosophy* (Humanities Press, Atlantic Highlands, New Jersey, 1977)

Findlay, J. N. 'Time: A Treatment of Some Puzzles', in Gale, *Philosophy of Time*, pp. 143–62, reprinted from *Australasian Journal of Philosophy* 19 (1941)

Flew, A. 'The Sources of Serialism', in S. C. Thakur (ed.), *Philosophy and Psychical Research* (George Allen and Unwin, London, 1976) pp. 81–96

Gale, R. M. 'Tensed Statements', *Philosophical Quarterly* 12, no. 46 (1962) pp. 53–9

——— 'A Reply to Smart, Mayo and Thalberg on "Tensed Statements"', *Philosophical Quarterly* 13 (1963) pp. 351–56

——— 'Some Metaphysical Statements About Time', *Journal of Philosophy* 60, no. 9 (1963) pp. 225–37

——— '"Is it Now Now?"', *Mind* 73, no. 289 (1964) pp. 97–105

——— *The Language of Time* (Routledge and Kegan Paul, London, 1968)

——— (ed.). *The Philosophy of Time* (Humanities Press, New Jersey; Harvester Press, Brighton, 1968)

Grünbaum, A. 'The Status of Temporal Becoming', in Gale, *Philosophy of Time*, pp. 322–54

King-Farlow, J. 'The Positive McTaggart on Time', *Philosophy* 49, no. 188 (1974) pp. 169–78

MacBeath, M. 'Mellor's Emeritus Headache', *Ratio* 25, no. 1 (1983) pp. 81–8

Mayo, B. 'Infinitive Verbs and Tensed Statements', *Philosophical Quarterly*, 13, no 53 (1963) pp. 289–97

McTaggart, J. M. E. 'Time', in Gale, *Philosophy of Time*, pp. 86–97, from J. M. E. McTaggart, *The Nature of Existence*, vol. 2 (Cambridge University Press, 1927) Book V, Chapter 33

Mellor, D. H. *Real Time* (Cambridge University Press, 1981)

——— "'Thank Goodness That's Over"', *Ratio* 23 (1981) pp. 20–30

——— 'MacBeath's Soluble Aspirin', *Ratio* 25, no. 1 (1983) pp. 89–92

Mink, L. O. 'Time, McTaggart and Pickwickian Language', *Philosophical Quarterly* 10, no. 40 (1960) pp. 252–63

Moore, G. E. *Some Main Problems of Philosophy* (George Allen and Unwin, London, 1953)

Mundle, C. W. K. 'Broad's Views About Time', in P. A. Schilpp (ed.), *The Philosophy of C. D. Broad* (Tudor, New York, 1959) Chapter 10, pp. 353–74

Nagel, T. 'Death', in T. Nagel, *Mortal Questions* (Cambridge University Press, 1979) pp. 1–10, reprinted from *Noûs* 4, no. 1 (1970)

Pears, D. F. 'Time, Truth, and Inference', in A. Flew (ed.), *Essays in Conceptual Analysis* (Macmillan, London, 1956) pp. 228–52

Prior, A. N. 'Time After Time', *Mind* 67 (1958) pp. 244–46

——— 'Thank Goodness That's Over', *Philosophy* 34 (1959) pp. 12–17

Romney, G. 'Temporal Points of View', *Proceedings of the Aristotelian Society* 78 (1977–78) pp. 237–52

Rorty, A. O. 'Fearing Death', *Philosophy* 58 (1983) pp. 175–88

Russell, B. *Principles of Mathematics* (Cambridge University Press, 1903)

——— 'On the Experience of Time', *Monist* 25 (1915) pp. 212–33

Ryle, G. 'It Was to Be', in G. Ryle, *Dilemmas* (Cambridge University Press, 1960) pp. 15–35

Schlesinger, G. N. *Aspects of Time* (Hackett Publishing Co., Indianopolis, 1980)

——— 'How Time Flies', *Mind* 91 (1982) pp. 501–23

——— *Metaphysics* (Basil Blackwell, Oxford, 1983)

Silverstein, H. S. 'The Evil of Death', *Journal of Philosophy* 77, no. 7 (1980) pp. 401–24

Smart, J. J. C. 'The Temporal Asymmetry of The World', *Analysis* 14, no. 4 (1954) pp. 79–83

——— 'The River of Time', in A. Flew (ed.), *Essays in Conceptual Analysis* (Macmillan, London, 1956) pp. 213–27, reprinted from *Mind* 58 (1949) pp. 483–94

——— ' "Tensed Statements": A Comment', *Philosophical Quarterly* 12 (1962) pp. 264–5

——— *Philosophy and Scientific Realism* (Routledge and Kegan Paul, London, 1963)

——— 'Spatialising Time', in Gale, *Philosophy of Time*, reprinted from *Mind* 64 (1955)

——— 'Time and Becoming', in P. van Inwagen (ed.), *Time and Cause* (D. Reidel, Boston and London, 1980) pp. 3–15

Smart, N. 'Philosophical Concepts of Death', in A. Toynbee *et al.*, *Man's Concern With Death* (Hodder and Stoughton, London, 1968) pp. 23–35

Sosa, E. 'The Status of Becoming: What is Happening Now?', *Journal of Philosophy* 76 (1979) pp. 26–42

Strawson, P. F. 'On Referring', in P. F. Strawson, *Logico-Linguistic Papers* (Methuen, London, 1971) pp. 1–27, reprinted from *Mind* 59 (1950); also in A. Flew (ed.), *Essays in Conceptual Analysis* (Macmillan, London, 1956)

Sumner, L. W. 'A Matter of Life and Death', *Noûs* 10 (1976) pp. 145–71

Taylor, R. 'Fatalism', in Gale, *Philosophy of Time*, pp. 221–31, reprinted from R. Taylor, *Metaphysics*, 1st edn (Prentice-Hall, Englewood Cliffs, New Jersey, 1963). Printed also as 'Fatalism' in *Philosophical Review* 71 (1964)

—— *Metaphysics*, 2nd edn Prentice-Hall, Englewood Cliffs, New Jersey, 1974)

Thalberg, I. 'Tenses and "Now"', *Philosophical Quarterly* 13 (1963) pp. 298–310

—— 'Fatalism Toward Past and Future', in P. van Inwagen (ed.), *Time and Cause* (D. Reidel, Boston and London, 1980) pp. 27–47

Toynbee, A. 'The Relation Between Life and Death, Living and Dying', in A. Toynbee *et al.*, *Man's Concern With Death* (Hodder and Stoughton, London, 1968) pp. 259–71

Whiteley, C. H. *Mind in Action* (Oxford University Press, 1973)

Williams, B. 'The Makropulos Case: Reflections on the Tedium of Immortality', in B. Williams, *Problems of the Self* (Cambridge University Press, 1973) pp. 82–100

Williams, D. C. 'The Myth of Passage', in Gale, *Philosophy of Time*, pp. 98–116; also in *Journal of Philosophy* 48 (1951) and C. C. Thomas (ed.), *Principles of Empirical Realism* (Springfield, Illinois, 1966)

Wittgenstein, L. *Philosophical Investigations* (Basil Blackwell, Oxford, 1953)

—— *Tractatus Logico-Philosophicus* (Routledge and Kegan Paul, 1961)

—— *The Blue and Brown Books* (Basil Blackwell, Oxford, 1969)

Index